T0146293

Investigation, Remediation and Protection
of
Land Resources

Investigation, Remediation and Protection of Land Resources

Dieter D. Genske

Whittles Publishing

CRC Press
Taylor & Francis Group

Published by
Whittles Publishing,
Dunbeath,
Caithness KW6 6EY,
Scotland, UK
www.whittlespublishing.com

Distributed in North America by
CRC Press LLC,
Taylor and Francis Group,
6000 Broken Sound Parkway NW, Suite 300,
Boca Raton, FL 33487, USA

ISBN 978-1870325-87-5
USA ISBN 978-1-4200-6021-8

Every effort has been made to correctly
acknowledge illustrative material. The publishers
ask that any omissions be reported to them
for incorporation in future editions.

The publisher assumes no responsibility for any injury and/or damage to persons or property from the
use or implementation of any methods, instructions, ideas or materials contained within this book.
All operations should be undertaken in accordance with existing legislation and recognised trade
practice. Whilst the information and advice in this book is believed to be true and accurate at the time
of going to press, the author and publisher accept no legal responsibility or liability for errors or
omissions that may be made.

Typeset by Compuscript Ltd., Shannon, Ireland

Printed in the UK by Athenæum Press Ltd., Gateshead

Contents

Preface

Land has always been the foundation of wealth and prosperity. It is the basis of 90 % of all human food. On land, plants grow which are needed to feed our livestock, build our houses and produce goods.

We have learned from the fading remnants of long gone cultures that the reason for their very disappearance has often been the unsustainable management of land resources. Today, we learn from scientific literature and documentaries that we are also destroying these resources, but on a global scale. If we fit together all the pieces of information, we can predict the state of our planet in five or ten generations from now. It appears likely that the land on which we live will be degraded to an extent so far unseen. Our planet, as lush and bountiful as it was, will have lost much of its hospitality due to the activities of its caretakers; it will have been degraded by us.

Stephen Hawking recently asked, "How can the human race survive the next hundred years?" In his public lecture *Life in the Universe*, Hawking answers this by pointing out that there would be "no time to wait for Darwinian evolution, to make us more intelligent, and better natured" and indicates that we are now "entering a new phase, of what might be called self designed evolution, in which we will be able to change and improve our DNA".

However, according to Ernst Ulrich von Weizsäcker of the Donald Bren School of Environmental Science and Management, Santa Barbara, our civilisation must survive the next hundred years by implementing *sustainable development*, which means utilising our limited resources much more efficiently.

Land is a basic and finite resource—and we are running out of it. Even today, only about a third of the total land surface remains *un*degraded, while a quarter is severely or very severely degraded. It is time to halt this reckless development. *Investigation, Remediation and Protection of Land Resources* aims to contribute towards this.

Dieter D. Genske
Berne, Switzerland

Acknowledgments

I would like to thank my colleagues and friends who contributed ideas, case files and illustrations. I am indebted to the reviewers for their comments and proposals. I would also like to thank Elaine Rowan for careful editing, Sue Steven and Annette Rettie for marketing and Keith Whittles for producing this book. And finally, I would like to thank the students who followed my courses in Germany, Japan, Switzerland, the Netherlands and South Africa. Their questions and critique have been the most precious source of inspiration.

1 Introduction

In 2004, the *Living Planet Report* of the Worldwide Fund for Nature (WWF) reported that humans consumed 20% more natural resources than the earth could produce, and that populations of terrestrial, freshwater and marine species fell on average by 40% between 1970 and 2000. In 2004, the average ecological footprint i.e. the amount of land needed to produce the resources required to sustain life and to take up wastes (Wackernagel and Rees 1996) has been assessed as 2.2 ha person^{-1}. However, there are only 1.8 ha of land available to provide natural resources for each of the people on this planet.

Land is the basis for 90% of all human food, livestock feed, fibre and fuel (EEA 2006). Land is a basic and finite resource—and we are running out of it. Even worse, we are degrading large amounts of it. In 1990, a global survey on the degree of degradation (Oldeman *et al.* 1991) revealed that of the total land surface (14 894 Mha), 13% (1964 Mha) is classified as degraded to some extent i.e. an area somewhat larger than the Russian Confederation. Of this area, 5% (749 Mha) is considered as lightly degraded, 6% (911 Mha) as moderately degraded, 2% (296 Mha) as strongly degraded and less than 0.1% (9 Mha) as extremely degraded due to human intervention (see Figure 1.1).

84% of human-induced degradation manifests as erosion (water and wind), 12% as chemical alteration and 4% as physical alteration. Of the human-induced degradation, 35% is caused by overgrazing, 30% by deforestation, 28% by bad farming techniques, 7% by overexploitation of vegetation for domestic use (firewood, timber, etc.) and 1% by industrialisation.

A criticism, however, is that the extent of actual degradation has been underestimated. For instance, the land surrounding a degraded terrain is also affected and thus downgraded. Furthermore, off-site effects such as sedimentation of eroded material need to be taken into account. In order to compensate for these shortcomings, the notion of severity of degradation was introduced (Bot *et al.* 2000). According to this approach, 90% of the long-settled lands of Europe are degraded to some degree. 58 countries, 21 of them in Europe, have no undegraded land left, that is, all land in these countries shows at least some degree of deterioration. 15 countries have more than 90% of their land severely degraded, another 17 countries have over 75% and a further 41 countries have over 60% of their land severely degraded. Only about one-third of the total land surface shows no signs of degradation, while one-quarter is severely or very severely degraded as indicated for different regions in Figure 1.2. It is likely that the economic damage due to land degradation is equivalent to at least 10% of the economic value of global agricultural production.

1

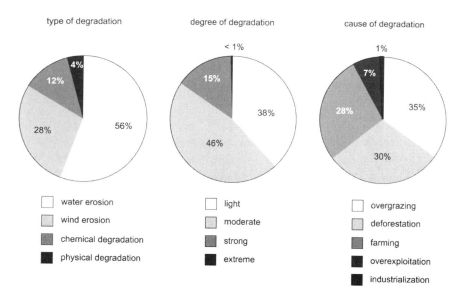

Figure 1.1 *Human-induced degradation of soil according to the type (erosion, chemical, physical), degree and cause of degradation (after Oldeman et al. 1990, FAO-AGL 2000)*

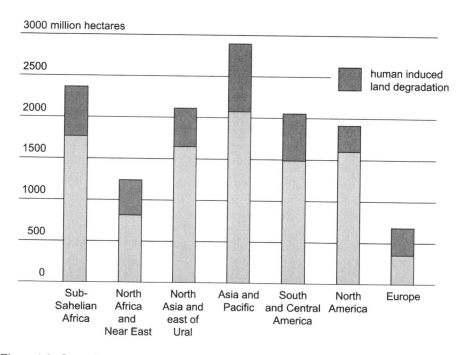

Figure 1.2 *Severely and very severely degraded land as a fraction of total land area (after Bot et al. 2000)*

It might well be that a closer investigation of the expenses associated with the remediation of degraded and contaminated industrial wasteland will significantly increase these costs, especially since large terrains of industrial degradation have not yet been assessed. In many European countries, not even half of the existing industrial wasteland has been screened and not even one-fifth has been investigated in detail. In some European countries, especially those located within the former Soviet Block, bases of the Warsaw Pact had occupied vast areas. In Estonia, Russian military bases had covered 1.8% of the land surface (EEA 2000).

There is a clear correlation between population density and human-induced soil degradation i.e. erosion plus chemical and physical degradation, as shown by Figure 1.3. With increasing population the quality of soil decreases and soil functions are lost. However, once land becomes increasingly degraded and eventually turns into sterile and toxic wasteland, it is abandoned and people are forced to move to other regions, accelerating the process of land degradation. The only measure to break this vicious circle is the introduction of sustainable land management techniques.

In 1992, the Earth Summit in Rio de Janeiro addressed the problem with its *Agenda 21*, stating that our practice of consuming resources shall not compromise on the ability of future generations to benefit from these resources, a principle that has become known as *sustainable development* (World Commission on Environment and Development 1987). In Chapter 10 of *Agenda 21* the resource land is recognised as a finite resource that shall be managed in a sustainable way. Land degradation should be stopped and measures of prevention and remediation shall be introduced.

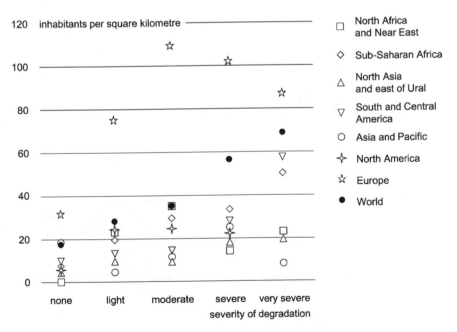

Figure 1.3 *Increase of land degradation with increasing population density (after Bot et al. 2000)*

Since then, the problem of land degradation has been analysed by an impressive number of agencies, organisations and individual scientists. Many articles, scientific papers and books have been published. The angle from which land degradation is addressed, however, varies depending on the parties involved and the goal set. When reducing the problem to a purely scientific analysis, land degradation offers aspects relevant to many disciplines including geo-, environmental, social and political sciences. These disciplines consist of a confusing number of sub-disciplines. Within the field of geosciences, for instance, engineering geology, hydrogeology, soil sciences, geo-informatics and geo-engineering all contribute knowledge to resolving aspects of land degradation. Within the field of environmental science, ecological sciences, environmental statistics, environmental engineering all contribute knowledge. In addition, disciplines like agriculture and regional planning are contributing to the fight of land degradation. All these fields and disciplines also consist of different specialisations and research domains. In addition, various traditions and schools have emerged in different countries like the Russian School of Soil Sciences opposing French or American equivalents. Sometimes, people concerned with soil conservation can be discussing the same parameter with different names. Sometimes parameters have the same name, but are measured in different ways. Sometimes the measuring method is comparable but the conclusions are different.

Due to the fragmentation of sciences into many disciplines, sub-disciplines and research schools, which are nevertheless bending over the same issue, working on land degradation and rehabilitation has become occasionally confusing, sometimes even discouraging. This is regrettable since all scientists involved are actually trying to present measures of land remediation and protection.

This book presents information from different scientific fields relevant to the complex problem of land degradation and rehabilitation. It brings them together to harmonise findings and suggest trans-disciplinary solutions. It attempts to do this in a straightforward language, understandable to everybody interested in the urgent task of sustainable land management.

This book is divided into three major parts:

- natural land
- degraded land
- remediating land.

The first part begins with an introduction into natural ground profiles, beginning with the perspective of soil sciences and concluding with a geological perspective of what is underneath the soil cover. In the following chapter, the functions of natural ground are discussed and the profound importance of ground for our society is outlined. Ground properties from an engineering point of view are described.

The second part of the book discusses degraded ground. Three mechanisms of ground degradation are introduced: erosion, chemical degradation and physical degradation. There are two aspects of erosion, notably water erosion and wind erosion. Four aspects of chemical degradation will be developed, notably contamination, salinization, acidification and solidification. Four aspects of physical degradation will be presented, notably surface sealing, compaction, subsidence and waterlogging.

The third and last part of this book discusses the remediation of degraded land. After introducing approaches and techniques to investigate ground and the grade of its deterioration, remediation measures are presented, focusing on erosion, chemical degradation and physical degradation. Finally, initiatives of land protection are outlined and a general strategy of sustainable land management is discussed.

In 1954, Rachel Carson who published *Silent Spring* stated:

"The more clearly we can focus our attention on the wonders and realities of the universe about us, the less taste we shall have for destruction. "

With this, she logically extended into our times Immanuel Kant's notion of

"...the starry sky above me and the moral law within me. "

that is, the obligation of humans to accept responsibility for the environment in which they live (Kant 1788). This responsibility shall be based on the moral concept that humans have the capacity to develop—if they dare to take advantage of their own intellect (Kant 1784).

It is hoped that this textbook will contribute to an understanding of the value of resource land, the way that humans degrade it and the options at hand to restore, rehabilitate and protect it.

I Natural Land

First, the development of land with its landforms is introduced and specific ground features are explained. The functions of natural ground are described and the benefits to our society presented. The properties of natural ground are explained and engineering aspects are developed.

2 Profiles

Ground basically consists of an upper active layer that is traditionally interpreted from the point of view of soil sciences as well as an underlying layer of parent material, traditionally interpreted through the eyes of the geologist. The mechanisms leading to soil and rock formation are discussed and typical soil and rock profiles are introduced. The major soil and rock groups are presented.

2.1 Pedosphere

2.1.1 Pedogenesis

Soil forms the *pedosphere* (Greek for soil) that separates the atmosphere from the *lithosphere* (Greek for rock). The French soil scientist Albert Demolon (1881–1954) describes soil as

> "a natural formation of the surface of the earth of loose structure and varying depth that results from the transformation of the exposed parent material due to physical, chemical and biological processes under the impact of the climate, the flora and the fauna".

With this definition Demolon pointed out the major factors of soil formation. The Russian pedologist Vasily V. Dokuchaev (1840–1903) puts these parameters into a simple formula:

$$S = f(P,\ C,\ O,\ R,\ T,\ A)$$

where S is soil, P the unweathered parent material, C the climate, O the organisms living on or within the soil, R the relief, T the time and A the anthropogenic impact. Thus, he recognises the activities of man as a soil forming—or soil degrading—factor. The development of soil may take only a few thousand years or up to tens of thousands of years. The degradation of soil, however, may take place quickly, sometimes within only a few centuries, decades or even years.

Besides Demolon and Dokuchaev, many more scientists contributed to the broad field of soil sciences including Justus von Liebig (1803–73), Curtis Fletcher Marbut (1863–1935), Hans Jenny (1899–1992) and Guy Smith, who introduced the US Soil Classification System in 1965.

For a soil profile to develop, a quasi-static environment is needed. Soils can only develop when the rate of soil formation exceeds the rate of erosion. This applies to both soils that develop on bedrock and soils that develop on sediments such as loess or glacial till that are themselves a result of weathering, erosion and sedimentation processes.

There are three different kinds of weathering:

- physical
- chemical and
- biological.

Physical (or *mechanical*) *weathering* includes the degrading effect of temperature, that is, the development of cracks and fissures due to temperature changes. In regions where the temperature contrast between day and night is high, rock weathers rapidly due to the thermal stresses that build up within exposed rock portions. In deserts, for example, where the diurnal temperature contrast is high, rock flakes chip off to form well-rounded outcrops, a process called exfoliation. As well as the temperature contrast, stresses develop once the water trapped in the cracks and pores freezes. In fact, the volume increase may be up to 9 % and thus every freeze-thaw cycle results in additional stresses and consequently disintegration. In dry climates, water in pores and fissures may evaporate during daytime prompting the growth of salt crystals, which also introduces stresses leading to the degradation of rock. Over time, the interaction of all these different types of physical weathering breaks the parent material into smaller and smaller pieces. If these are not eroded away they constitute the material for a young soil or leptosol.

Chemical weathering is governed by the presence of water. In a complex reaction, feldspar ($K\,Al\,Si_3\,O_8$), abundantly present in many rocks such as granite, decomposes to kaolinite ($Al_2\,Si_2\,O_5\,(OH)_4$), a common clay mineral (Press *et al.* 2003)

$$2K\,Al\,Si_3O_8 + 2H_2CO_3 + H_2O \rightarrow Al_2Si_2O_5(OH)_4 + 4SiO_2 + 2K^+ + 2H\,CO_3^-$$

Similar types of reactions can be observed for other silicates. As can be seen from this reaction, carbonic acid is needed to form kaolinite. It is usually present in the rainwater that comes into contact with the outcropping rock. However, as soon as soil covers the rock, the process of weathering accelerates since the percolating rainwater picks up additional acids produced by the roots of plants and other organisms living in the soil. As a result, rock overlain by a soil cover weathers much faster. It also weathers steadily since percolating water is trapped within the soil matrix. The reactions that form clay minerals are generally referred to as hydration since they consume water. Chemical weathering however, embraces many more types of reactions such as oxidation processes that break down iron silicates to ferric iron oxides, which give the soil the characteristic yellowish-brownish colour. While the young soil matures, dissolved minerals, iron and aluminium oxides, clay particles and organic matter migrate with percolating rainwater to accumulate in a deeper soil horizon, a process referred to as illuvation.

Biological weathering is caused by soil organisms which contribute acids and other chemicals to break down the parent material. Furthermore, plant roots and soil animals loosen the soil in such a way that water and oxygen can easily reach rock fragments to further break them down. In addition, decaying organic matter produces humic acid that further stimulates chemical weathering processes.

The different types of weathering complement and enhance each other. They accelerate the disintegration and decay of the parent material. In general, it can be stated that the weathering rate increases with rainfall, temperature contrasts, mineral solubility, soil cover, flora and fauna. The interaction of all these factors makes the process of soil development or pedogenesis a manifold and complex one. Many of the phenomena introduced in this chapter have been discovered only recently while others are still to be investigated. Indeed, as Leonardo da Vinci (1452–1519) stated:

"We know more about the movement of celestial bodies than about the soil under our feet."

The efforts to introduce soil classification systems are almost as abundant as the types of soils that can be distinguished. For example, the Russian classification system is dominated by the key parameters climate and vegetation, whereas the scheme developed in North America is directed towards soil horizons and their chemical and physical properties. A similar system has been proposed by the Food and Agriculture Organisation (FOA) of the United Nations, while the French classification focuses on the natural development of the soil profile.

An initial approach to classifying soil types is to distinguish between

- soils that have developed from the bedrock on which they lie and
- soils that have developed on transported sediments and therefore not related to the bedrock.

In the former case, the type of bedrock present at the site can usually be inferred from the soil profile, whereas in the latter case the soil profile only indicates the type of sediment that covers the bedrock (Figure 2.1).

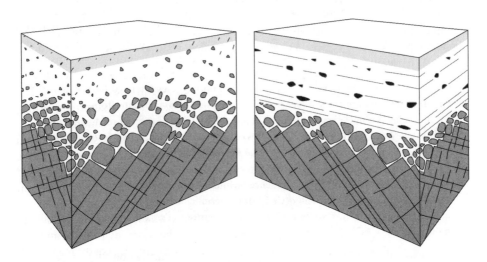

Figure 2.1 *Certain soils develop from the bedrock on which they lie, whereas others develop on transported sediments*

Another distinction can be made between

- residual soils, which develop in place either on bedrock or sediments and represent most of the soil types, and
- transported soils, which owe their profile to the deposition of eroded soil in depressions and lowlands.

Another approach to grouping soil types is to distinguish between

- natural soils and
- anthropogenic soils,

the latter of which can further be divided into

- agricultural soils
- technical soils, for example fills, heaps and tailings, and
- urban soils, also referred to as urbic anthrosols.

In the 1980s, the FAO and the UNESCO started to think about a common taxonomy to define the major soil groups, supported by United Nations Environmental Programme UNEP and the International Society of Soil Sciences ISSS. The final goal was to map all major soils at a global scale. Later, the International Soil Reference and Information Centre ISRIC, emerging from the International Soil Museum in Wageningen (Netherlands), became involved in the project. The result was the *World Reference Base* (WRB) that introduced internationally harmonised soil groups (IUSS 2006). A first version of the WRB Soil Resource Map was completed in 1990 at a scale 1:25 000 000 (1 cm $\equiv$ 250 km). The map has been frequently updated and initiatives have been launched to create world soil and terrain databases (e.g. SOTER, Baumgardner 2000, Engelen 2000). The reference soil groups that were also adopted for maps at scales relevant for practical engineering works are explained in Table 2.1.

The name of each reference soil group may be further modified by linguistic qualifiers. The IUSS (2006) distinguishes prefix qualifiers i.e. typically associated and intergrade qualifiers as well as suffix qualifiers indicating diagnostic horizons, chemical, physical and mineralogical characteristics, textural characteristics and soil colour. For instance (IUSS 2006: 21)

"A soil has a ferralic horizon [i.e. a subsurface horizon characterized by Fe- and Al- (hyrd)oxides]; texture in the upper part of the ferralic horizon changes from sandy loam to sandy clay within 15 cm. The pH is between 5.5 and 6, indicating moderate to high base saturation. The B-horizon [Chapter 3] is dark red; below 50 cm, mottling occurs. The field classification of this soil is: *Lixic Ferrasol (Ferric, Rhodic)* [Lixic indicating a subsurface horizon with high clay content, Ferric indicating a segregation of Fe, Rhodic indicating the red colour]. If subsequent laboratory analysis [Chapter 4] reveals that the cation exchange capacity (CEC) of the ferralic horizon is less than 4 cmol kg^{-1} clay, the soil finally classifies as Lixic Vetic Ferrasol (Ferric, Rhodic) [with Vetic indicating low CEC]."

This example demonstrates the complexity of soil identification. In fact, profound background knowledge and substantial field experience is needed to correctly identify soil types.

Table 2.1 Soils of the World Reference Base for Soil Resources WRB (after ISSS Working Group 1998, Zech and Hintermaier-Erhard 2002, IUSS 2006)

RSG[1]	Acronym	Origin[2]	Typical Climate[3]	Typical Profile	Description
Acrisols	AC	acer [La] = acid	(E), G, **H**, I	AEBtC	Soils with subsurface accumulation of low activity clays and low base saturation
Albeluvisols	AB	albus [La] = white, eluere [La] = leach	**B**, C	(O)AE(g)Bt(g)C	Acid soils with a bleached horizon penetrating into a clay-rich subsurface horizon
Alisols	AL	alumen [La] = aluminium	(C, E), G, **H**, (I)	AEBtC	Soils with subsurface accumulation of high activity clays, rich in exchangeable aluminium
Andosols	AN	an, do [Ja] = black, soil	(B, C, D, E, F, G, H, I), **J**, K	AC, ABC	Young soils from volcanic deposits
Anthrosols	AT	anthropos [Gr] = man	**K**	Ap(B)C, ApgBgC	Soils in which human activities have resulted in profound modification of their properties
Arenosols	AR	arena [La] = sand	D, **F**, G, (H)	AC, AEC	Sandy soils featuring very weak or no soil development
Calcisols	CL	calx [La] = lime	D, (E), **F**, (G, J)	AC(c)kC, AB(e)wkC, AB(e)tkC	Soils with accumulation of secondary calcium carbonates
Cambisols	CM	cambiare [It] = change	(A, B), **C**, (D), **E**, F, (G, H, I), J, K	ABwC	Weakly to moderately developed soils
Chernozems	CH	tschornyi [Ru] = black, zemlja [Ru] = ground	C, **D**, (J)	AhC(c)kC	Soils with a thick, dark topsoil, rich in organic matter with a calcareous subsoil
Cryosols	CR	kryos [Gr] = cold, ice	**A**, B, J	ABCf, ABfCf, ACf	Soils with permafrost within 1 m depth
Durisols	DU	durus [La] = tough	(D), **F**	AC(m)q, AB(m)qC	Soils with accumulation of secondary silica
Ferralsols	FR	ferrum, alumen [La] = iron, aluminium	(F), G, H, **I**	ABwsC	Deep, strongly weathered soils with a chemically poor, but physically stable subsoil

13

Table 2.1 (continued)

RSG¹	Acronym	Origin²	Typical Climate³	Typical Profile	Description
Fluvisols	FL	fluvius [La] = river	B, C, D, E, F, G, H, I, **K**	ACg2Cg3Cg, AC2AlBw3Cg	Young soils in alluvial deposits
Gleysols	GL	gley [Ru] = wet loamy soil	A, **B**, C, (D, E), G, H, I, K	ACr, ABgCr, HCr, HBgCr	Soils with permanent or temporary wetness near the surface
Gypsisols	GY	gypsum [La] = gypsum	D, (E), **F**	AC, ABwyC, ABtyC	Soils with accumulation of secondary gypsum
Histosols	HS	histos [Gr] = fabric	A, **B**, (C, G), I, (K)	HCr	Soils which are composed of organic materials
Kastanozems	KS	castaneo [La] = chestnut, zemlja [Ru] = ground	**D**, (E), F, (G), J	AhC(c)kC(y), AhB(c)kC(y)	Soils with a thick, dark brown topsoil, rich in organic matter and a calcareous or gypsum-rich subsoil
Leptosols	LP	leptos [Gr] = slim	A, B, (C, D), E, **F**, (G, I), **J**, K	A(B)C, A(B)R	Very shallow soils over hard rock or in unconsolidated very gravelly material
Lixisols	LX	lixivia [La] = washed out substrate	(F), **G**, (H, I)	AEBtC	Soils with subsurface accumulation of low activity clays and high base saturation
Luvisols	LV	eluere [La] = leach	(B), **C**, D, **E**, F, (G), H, (J)	AEBtC	Soils with subsurface accumulation of high activity clays and high base saturation
Nitisols	NT	nitidus [La] = shining	(F), **G**, H, I	A(E)BtC	Deep, dark red, brown or yellow clayey soils having a pronounced shiny, nut-shaped structure
Phaeozems	PH	phaios [Gr] = dim, zemlja [Ru] = ground	(B), C, **D**, H, (I, J)	AhBwC, Ah(E)BtC	Soils with a thick, dark topsoil rich in organic matter and evidence of removal of carbonates
Planosols	PL	planus [La] = flat	C, D, (E, F), **G**, (H)	AEBgC	Soils with a bleached, temporarily water-saturated topsoil on a slowly permeable subsoil

			Eco-zone[3]	Profiles	Description
Plinthosols	PT	plinthos [Gr] = brick	G, H, **I**	AB(m)sqC, AEB(m)sqC	Wet soils with an irreversibly hardening mixture of iron, clay and quartz in the subsoil
Podzols	PZ	pod [Ru] = below, zol [Ru] = ashes	**B**, C, (G, H), I, J	OAEBhsC	Acid soils with a subsurface accumulation of iron-aluminium-organic compounds
Regosols	RG	rhegos [Gr] = cover	A, (B, D), **F**, (G, H), **J**, K	AC	Soils with very limited soil development
Solonchaks	SC	sol [Ru] = salt	(A), D, (E), **F**, (I)	Az(Bzg)Czg, A(Bzg)Czg	Strongly saline soils
Solonetz	SN	solonez [Ru] = salt water	(B, C), **D**, (E), F	ABtnC, AEBtnC	Soils with subsurface clay accumulation, rich in sodium
Stagnosols	ST	stagnare [La] = to flood	B, C, H	AC, AbwC, AEBtC	Periodically wet soils, also known as pseudogley
Technosols	TC	technikos [Gr] = skilfully made	K	none or only little profile development	Soils dominated by human made materials like wastes, mine spoil, sludge, cinder and ashes
Umbrisols	UM	umbra [La] = shadow	B, C, (E, F), G, (H), I, **J**	AC, AEC, ABC	Acid soils with a thick, dark topsoil rich in organic matter
Vertisols	VR	vertere [La] = turn	(C, D, E), F, **G**, (H)	AC, ABC	Dark-coloured cracking and swelling clays

[1] Reference Soil Group

[2] Gr Greek, It Italian, Ja Japanese, La Latin, Ru Russian

[3] A polar and subpolar zones (tundra), B boreal zone (taiga), C wet mid-latitude zone (temperate), D dry mid-latitude zone (steppe), E wet-winter subtropics (Mediterranean), F dry tropics and subtropics (deserts), G wet-dry tropical zone (savannah), H wet subtropical zone (eastern), I wet tropical zone (rain forest), J alpine regions, K generally occurring soils. **A** dominant soil group, A accompanying soil group, (A) sporadically occurring soil group (frequently at the border to other eco-zones).

Zech and Hintermaier-Erhard (2002) attempt to categorise the reference soil groups with regard to ecologically homogeneous zones. For every eco-zone they indicate dominant soil groups:

- Polar and subpolar zones (tundra) dominated by *cryosols* (north Alaska, north Canada, north Scandinavia, north Russia, coasts of Greenland and Iceland, higher Rocky Mountains, mountains of middle Siberia and northeast Siberia, Falkland Islands, coast of Antarctica)
- Boreal zone (taiga) dominated by *histosols, gleysols, stagnosols, podzols* and *albeluvisols* (Alaska, Canada, Scandinavia, Russia, as well as isolated regions in the Rocky Mountains, the Alps, Iceland, Carpathian Mountains, the Caucasus, the Tien Shan and the Himalaya)
- Wet mid-latitude zone (temperate) dominated by *cambisols, luvisols, umbrisols* and *stagnosols* (east and northeast USA and bordering Canada, southwest Canada and northwest USA, large parts of Europe, northeast and northwest China, Korea, north Japan, south Chile, southeast Australia, Tasmania, south New Zealand)
- Dry mid-latitude zone (steppe) dominated by *phaeozems, chernozems, kastanozems* and *solonetzes* (midwest USA up to south Canada, Central Asia, Patagonia)
- Wet-winter subtropics (Mediterranean) dominated by *chromic cambisols* and *chromic luvisols* (Mediterranean, California, northwest Pakistan, middle Chile, Cape of South Africa, southwest and south Australia)
- Dry tropics and subtropics (deserts) dominated by *arenosols, calcisols, gypsisols, solonchaks* and *durisols* (north Mexico, Sahara Desert, Horn of Africa, Arabian Peninsula, Mesopotamia, Iran, Afghanistan, Pakistan, northwest India, coast of Peru, central Andes, Atacama Desert, Gran Chaco, northeast Brazil, southwest Africa, Kalahari Desert and Karoo, deserts of central Asia)
- Wet-dry tropical zone (savannah) dominated by *lixisols, nitisols, vertisols* and *planosols* (highlands of Central America, Yucatan, west Cuba, north South America, parts of west, central and east Africa, southwest, south, southeast and east India, east Sri Lanka, west Philippines, central South America, major parts of west, central and east Africa, Madagascar Mountains, north and northeast Australia)
- Wet subtropical zone (eastern) dominated by *acrisols, alisols* and *stagnosols* (southeast USA, central and east China, south Korea, south Japan, southeast South America with south Brazil, the Pampa of northeast Argentina and Uruguay, southeast South Africa, east Australia and the northern island of New Zealand)
- Wet tropical zone (rain forest) dominated by *ferrasols* and *plinthosols* (east Central America, parts of the Caribbean Islands, northwest South America, coastal mountains of west Africa, northeast Congo, east India, Island of southeast Asia, Amazon, east South America, east coast of Madagascar).

The dominant reference soil groups are complemented with accompanying soil groups and sporadically occurring soil groups as listed in Table 2.1. Furthermore, soils such as

- *leptosols* (shallow soils), *regosols* (initial soils) and *andosols* (from volcanic ashes)

16

- *fluvisols* (river laid)
- *anthrosols* (distorted due to human activities) and
- *technosols* (soils of technical origin like landfills, sludge, cinders, mine spoils and ashes)

may occur in all eco-zones.

2.1.2 Soil profiles

When mapping soil, one will notice that different soil profiles can be distinguished. Upon closer investigation it is possible to recognise individual horizons that have developed over the parent material. A typical profile of a residual soil is composed of three distinct layers or master horizons:

- the upper A-horizon consists of topsoil rich in organic matter, clay and insoluble minerals
- B-horizon or subsoil, an accumulation layer consisting of soluble minerals and iron oxides (leached from the A-horizon)
- C-horizon representing the parent material

as shown in Figure 2.2.

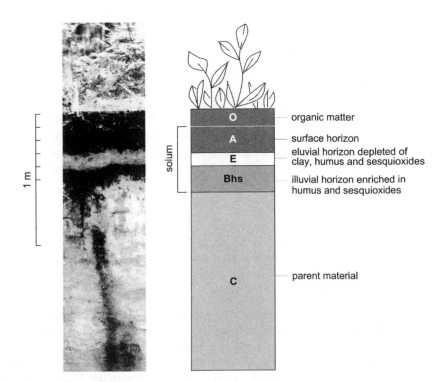

Figure 2.2 *The soil horizons of a podzol (late Pleistocene) from northern Germany (after Zech and Hintermaier-Erhard 2002)*

The A- and B-horizon are also referred to as *solum* (Latin for soil), the generic soil due to soil forming processes. Most chemical and biological processes take place within the solum. The C-horizon is the parent material uninfluenced by pedogenic processes, which may either be bedrock or a sediment layer. Sediments are grouped according to their grain size distribution as

- boulders (grain size > 200 mm)
- cobbles (grain size 60–200 mm)
- gravel (grain size 2–60 mm)
- sand (grain size 0.06–2 mm)
- silts (grain size 0.002–0.06 mm)
- clays (grain size < 0.002 mm) and
- mixtures of these groups such as silty sand or loam.

The type of sediment is determined by means of grading curves, explained in Chapter 4. Soils are usually autochton and thus derived directly from the parent material. Allochton components may affect the soil profile via runoff and flooding.

With the separation of the soil profile into A-, B- and C-horizons a straightforward terminology is introduced, supplemented by additional master horizons, notably:

- L-horizon, the surface cover of litter (decomposing leafs, twigs, etc.)
- O-horizon, rich in organic matter
- E-horizon, from which silicate clay, iron and aluminium are leached.

A general sequence of master horizons is written as L–O–A–E–B–C. Certain taxonomic systems feature the R-horizon as the unaffected bedrock underlying the C-horizon. Transitional horizons are annotated as a combination of the master symbols. For instance, EB denotes a horizon that exposes properties of both the E-horizon and the B-horizon. Since E is mentioned first, the EB-horizon is dominated by E-features.

Added to the capital symbols are subordinate small letters, which further characterise the properties of the horizons. Given the nomenclature for the master horizons, the transitional horizons and the subordinate features, the description of a soil profile can become rather complex. In addition to the aspects already mentioned, more data are collected describing the

- texture of the soil: the proportion of sand, silt and clay
- structure of the soil: the amount and distribution of pores and soil particles
- acidity of the soil: its pH-value (typically between 4–8)
- soil colour as defined by the *Munsell Book of Colours* with huc, value and chroma
- degree and distribution of mottling.

Given the same climatic conditions, typical soil profiles develop depending on the parent material and the topography. In order to illustrate the spatial development of soil profiles, pedologists draw *catenas* (chains) or causal profiles illustrating the possible manifestations of soil development, as shown in Figure 2.3 These catenas can be considered as simplified illustrations of how soil profiles develop depending on their location within a given topography. They are quite helpful when mapping soil since they indicate the types and variations of soils to be expected.

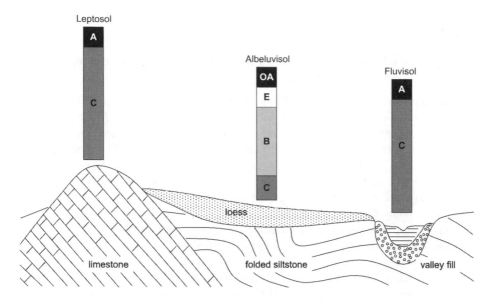

Figure 2.3 *A catena with three typical soil profiles (simplified)*

Substantial experience is needed in order to properly map a soil profile, define its horizons and derive the generic type and origin of the soil under investigation. It is not the task of the engineer to undertake this work. However, when discussing soil degradation and remediation problems it is certainly an advantage to be familiar with the terminology of the soil scientist in order to avoid misunderstandings and to harmonise the work.

2.2 Below the pedosphere

2.2.1 Sediments and rocks

Although valleys and mountains appear to be static, they are actually the result of dynamic processes that constantly model and form the earth's surface. The extent of these changes is not perceptible to the human eye over long periods of time. However, it was as early as 500 BC that the Greek historian Herodot recognised that the Nile Delta gradually changes. Two thousand years later, Giordano Bruno (1548–1600) wrote (Blei 1981) that

"…there soon is the sea where once was a river, there soon are mountains, where once were valleys…But in all this I cannot see any violence but only a natural development. For I see only violence in what happens outside the limits of nature or even against it. "

For Giordano Bruno this conclusion and other findings that appear common to us today led straight to the stakes of the holy inquisition, where he was burned on the 17th of February 1600. Despite this, the Danish medical doctor, theologian and natural scientist Niels Stensen (1638–86) stated in his *De solido intra solidum naturaliter contento dissertationis prodomus* that the former state can be read for the present state, meaning that geological features of today give hints of how they developed in the past. Deciphering the code to reveal the past to explain the present is the very task of the geologist.

About a hundred years later, Abraham Gottlob Werner (1749–1817) who taught at the newly established Bergakademie in Freiberg, Germany, noted (Hölder 1989)

"…that the solid surface of the earth is a child of time, a result of natural changes […] that must be studied by the geognost."

Georg Friedrich Füchsel (1722–73) agreed with this point of view in his *Outline of the oldest history of earth and man* (1773) and the British geologist James Hutton (1726–97), Figure 2.4, recognised in this statement the basic idea of explaining geological features. His colleague Charles Lyell (1797–1875) introduced the term *actualism* that he explained in more detail in his *Principles of Geology* (1830–33). For instance, he interpreted rightly that the shape of the Island of Santorin resembles an extinct volcano, thus the island must have been formed by volcanic activities. In the same way, Otto Martin Torell (1828–1900) concluded rightly that the peculiar parallel scratches in the Rüdersdorfer Mountains close to Berlin must have been caused by glaciers that once covered the whole of Northern Europe. With this, the director of the Swedish National Office of Geology and Professor of Zoology at the University of Lund confirmed the supposition of the German mineralogist Johann Friedrich Ludwig Hausmann (1782–1859) that the 'erratic blocks' found in Northern Germany originated from Scandinavia—an assumption that was incidentally disputed by Charles Lyell. Actualism allows, then and today, an interpretation of nature as a book with the surface of the earth as text that is to be interpreted *per analogiam* (Genske and Hess-Lüttich 1998). In order to sort the plentiful fragments of information and to derive reliable theories about the formation of geological features, the facies have to be read in the right way, *mente et malleo*, with brain and hammer.

Today, this concept is still topical, given that an agreement has not been reached on explaining the origin of the earth. The English archbishop James Usher

Figure 2.4   *James Hutton reads the 'facies' of an outcrop (lithography from John Kays Portraits, Edinburgh 1837)*

(1581–1656) dated the origin of the earth as the night between the 22nd and 23rd October of 4004 BC, whereas Johann Albrecht Bengel (1687–1752) in Germany dated this event as the 10th of October 3943 BC. And yet, there is still disagreement. When in 1993 the German newspaper *Frankfurter Rundschau* discussed the extinction of dinosaurs millions of years ago, Rabbi Swi Gafner protested because "it is a commonly acknowledged fact that the world was created only 5753 years ago". In addition, much attention is today given to those proclaiming a creator who is patiently trying to design intelligent forms of life, offering explanations for everyone who resorts to given answers when subjects reach a certain degree of complexity.

Actualism has inspired geological thinking until today. Since Alfred Wegener (1880–1930) developed the continental drift theory it is known that forces underneath the earth's crust move tectonic plates that may drift apart or collide. These mechanisms are referred to as endogenetic dynamics. Along the collision zones mountain chains build up, a process known as *orogenesis*, while earthquakes release the stresses and volcanoes erupt. Water, wind and ice erode the mountains (see Figure 2.5) and transport the eroded material to places of lower relief energy where sediments accumulate to form new strata. These mechanisms of exogenetic dynamics are fundamental for the creation of landforms. On one hand, bedrock related landforms develop where the action of weathering and erosion alters and shapes the appearance of the rock masses. For example, the eroding action of water creates gullies and valleys that extend headward into higher land. The sandblasting action of wind creates *ventifacts* or wind-faceted stones such as *driekanter*. Glaciers shape horns with *arêtes* and *cirques* and U-shaped valleys that are accompanied by hanging valleys as well as polished rock surfaces and features like *roche moutonnée* (shaped bedrock hills). On the other hand, processes of weathering, erosion and sedimentation yield another large group of specific landforms (see Figure 2.6) that can be grouped into three classes:

- fluviatile
- aeolian or wind-laid and
- glacial.

Fluviatile landforms (Figure 2.7) include alluvial fans, valley fills, flood plains, terraces, coastal plains and deltas. Depending on the grain size, sediments are moved as fluvial bedload (by sliding and rolling), suspended load (suspended in the flow itself) or by saltation (by bouncing and touching the bed only momentarily). These sediments are eventually deposited as the flow velocity decreases.

Aeolian landforms (Figure 2.8) include sand dunes and loess deposits. Depending on the grain size, dust and sand particles are lifted and transported by the wind, a process known as deflation. Loess may be deposited far away from its source since the size of the particles ranges from only 0.01–0.05 mm, which refers to the silt fraction (0.002–0.06 mm). Sand particles are mainly moved by saltation, rolling and sliding to form barchan, transverse, linear and blowout dunes. This process is dependent upon the prevailing winds and the amount of sand available.

Glacial landforms (Figure 2.9) include alpine varieties and landforms created by continental glaciers that pick up and slice off rock and soil debris while they are moving. They deposit this material as ground moraine, a heterogeneous layer that develops underneath a glacier also referred to as glacial till consisting of clay, silt,

21

(a)

(b)

(c)

Figure 2.5 *Landforms created by the eroding action of water, wind and ice: (a) V-shaped valley formed by the Vorderrhein-River, Canton Graubünden, Switzerland, (b) Pancake erosion shaped by the action of the wind, Andalusia, Spain and (c) U-shaped valley carved by an alpine glacier, Maderanertal, Canton Uri, Switzerland*

Figure 2.6 *Landforms created by fluvial, eolian and glacial deposition: (a) braided river, New Zealand (photo courtesy of Monika Huch), (b) sand dunes, Namib Desert, Namibia and (c) glacial moraines, Scandinavia (photograph courtesy of Helmold Strübing)*

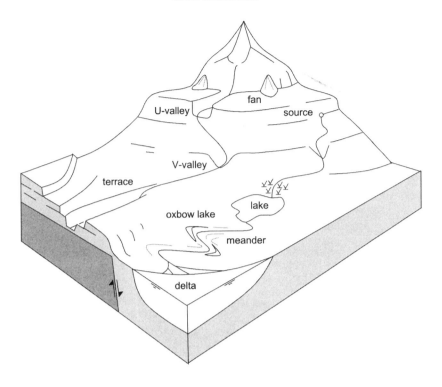

Figure 2.7 *Fluvial landforms (simplified)*

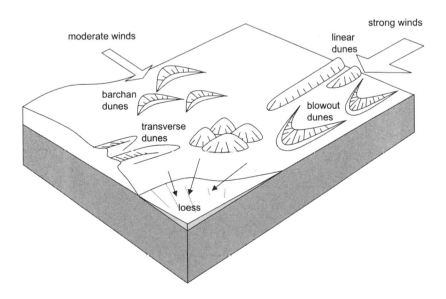

Figure 2.8 *Aeolian landforms (simplified)*

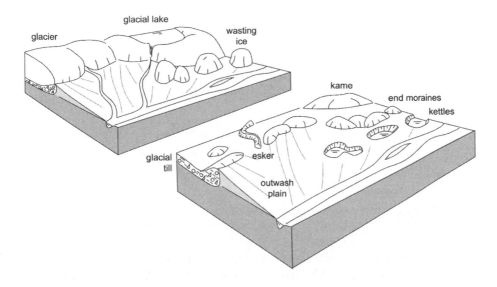

Figure 2.9 *Glacial landforms (simplified)*

sand, gravel, cobble, and boulders of a considerable size. When glaciers retreat, they leave end moraines as well as a variety of glaciofluvial landforms such as outwash plains, eskers (dam-shaped remnants of the bedload of streams flowing in, on or underneath a glacier) and kames (conical hills of sediments that accumulate in depressions on the surface of the glacier).

In the course of geological history, the level of the sea changes frequently. Vast land areas are flooded and receive enormous amounts of sediments. When the thickness of the sediment layer increases, pressure and temperature increase as well. The sediments are compacted and cemented by precipitating minerals while the porosity decreases, a process referred to as *diagenesis*. Eventually sedimentary rock develops, consisting of compacted sediments such as gravel, sand, silt and clay forming conglomerate, sandstone, siltstone and claystone as well as chemical sediments (see Tables 2.2 and 2.3). They record the chemical environment during the rock forming process such as gypsum or salt, or biochemical sediments such as limestone consisting of shells of aquatic organisms.

With increasing pressure and temperature metamorphic rock forms, as illustrated in Figure 2.10, while new minerals are generated from vanishing old ones. Typical examples

Table 2.2 *Typical clastic sediments and sedimentary rocks*

Sediment	Particle size (mm)	Rock
Boulders, cobbles, gravel	> 2	Conglomerate
Sand	0.06–2	Sandstone
Silt, clay	< 0.06	Siltstone, mudstone, shale

Table 2.3 *Typical chemical and biochemical sedimentary rocks and main components (after Press et al. 2003)*

Sediment	Principal component	Minerals	Rock
Carbonate sand and mud	$CaCO_3$	Calcite	Limestone
No primary sediment	$CaMg(CO_3)_2$	Dolomite	Dolostone
Iron oxide sediment	FeO_3, $FeCO_3$	Hematite, limonite, siderite	Iron formation
Evaporite sediment	NaCl, $CaSO_4$	Gypsum, anhydrite, halite, etc.	Evaporite
Siliceous sediment	SiO_2	Opal, chalcedony, quartz	Chert
Peat, organic matter	C	-	Coal, (oil, gas)
No primary sediment	$Ca_3(PO_4)_2$	Apatite	Phosphorite

are gneiss (derived from granite), quartzite (derived from sandstone) and marble (derived from limestone). See Table 2.4 for other examples. Regional metamorphism is typical for sinking basins into which sediments are deposited steadily. Metamorphic rock also develops along the collision zones of tectonic plates where one plate is pushed underneath the other. Within the subduction zone, temperature and pressure increases, thus creating the milieu of metamorphic rock to develop. Metamorphism is also a distinctive aspect in mountain building processes and vast resources of metamorphic rocks can usually be found in the core of a mountain zone. Moreover, metamorphism takes place along the contact to magma bodies where rock is still liquid (contact metamorphism). Even the

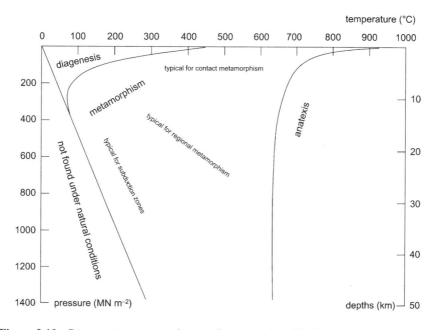

Figure 2.10 *Diagenesis, metamorphism and anatexis (simplified)*

26

Table 2.4 *Typical metamorphic rocks and parent rock from which the metamorphic variety was formed*

Parent rock	Metamorphic varieties
Sandstone	Quartzite
Shale, sandstone, greywacke	Slate, phyllite, schist, gneiss
Limestone, dolomite	Marble
Coal	Graphite
Granite	Gneiss
Basalt	Greenstone

impact of meteorites may produce metamorphic rock as for instance at the Nördlinger Ries, Germany, where the shock dynamics can be reconstructed by means of special minerals that have formed at the moment of the impact. For all scenarios, characteristic metamorphic facies can be distinguished.

With increasing temperature rock begins to melt, first partially and then completely, a process called *anatexis*. It mixes with molten rock or magma deep below the surface. When magma cools off and solidifies, igneous rock is formed. The process of cooling-off may take place gradually, underneath the surface of the earth for example, in which case intrusive igneous rock is formed (Figures 2.10 and 2.11).

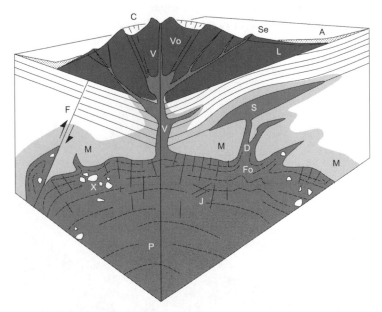

Figure 2.11 *Intrusive and extrusive igneous rock. A: ashes and pyroclastics, C: caldera, D: dike, F: fault, Fo: foliation, J: joint, L: lava, M: contact metamorphism, P: pluton (former magma reservoir), S: sill, Se: secondary volcano, V: volcanic conduit and central vent, Vo: volcano, X: xenolith (inclusions)*

Table 2.5 *Igneous rocks and their properties*

Intrusive igneous rocks	*Extrusive igneous rocks*	*Silicate content*	*Viscosity*	*Chemical composition*
Granite	Rhyolite	High	High	Felsic
Granodiorite	Dacite	Intermediate	Intermediate	Intermediate
Diorite	Andesite	Intermediate	Intermediate	Intermediate
Gabbro	Basalt	Low	Low	Mafic

Due to the long cooling time the minerals can grow to form a coarse crystalline texture with quartz filling in the gaps between the minerals. Depending on the chemical composition different varieties develop, from granite having a *felsic* composition (meaning rich in feldspar and silica) to Gabbro having a *mafic* composition (meaning rich in magnesium and ferrum i.e. iron) (see Table 2.5).

If magma rises to the surface it forms extrusive igneous rock or volcanic rock (Figures 2.11 and 2.12). As volcanic rock cools rapidly, crystals can barely grow and a finely crystalline texture develops. Different varieties can be distinguished, from rhyolite having a felsic composition to basalt having a mafic composition (Table 2.5). The eruption of volcanoes produces lava that may be fast flowing *pahoehoe* lava (mafic) or slowly moving *a-a* lava (felsic). As well as lava, a volcano also emits pyroclastics i.e. ashes, tuffs and breccias, which may roll downhill as pyroclastic flows at speeds of up to 200 km hr^{-1} or rain down to devastate whole regions such as Pompeii and Herculaneum at the foot of Mount Vesuvius in Roman times. Figure 2.12 depicts an example of a pyroclastic flow descending Unzen in Japan. Along the mid-ocean ridges where tectonic plates drift apart, submarine volcanoes produce pillow lava (lava blocks resembling pillows) due to the specific conditions at the sea floor.

Figure 2.12 *Pyroclastic flow descending the Unzen Volcano in Kyushu, Japan, February 2nd 1992 (photo courtesy of Kenji Ohkawa)*

Diagenensis, metamorphism and anatexis are fundamental rock forming mechanisms. A large and diverse spectrum of rock types develops from these processes. Their understanding is essential when ground conditions need to be investigated. An introduction into rock forming processes and rock types is given Press *et al.* (2003), for example.

Figure 2.13 shows typical examples of igneous, sedimentary and metamorphic rocks.

(a)

(b)

(c)

Figure 2.13 *Samples of (a) igneous; (b) sedimentary (photo courtesy of Franz Tessensohn) and (c) metamorphic rock*

2.2.2 Rock profiles

During mountain building, geological units are faulted and folded repetitively, leaving complex profiles difficult to interpret even for experienced field geologists. Many aspects can be misinterpreted and many hints can be misunderstood as the following example demonstrates (Trümpy 1991).

Hans Conrad Escher von der Lindt (1767–1823), a Swiss artist of profound geological knowledge, painted a number of impressive aquarelles depicting the alpine landscape, always trying to understand the inner structure of those imposing mountains. His son Arnold shared his passion and began to draw numerous panoramas and profiles. Later, he was appointed the first chair of geology at the Swiss Federal Institute of Technology, the ETH Zürich. During his many excursions he noted an interesting feature in the Canton of Glarus where older rock overlies younger formations, a phenomenon that he could only explain with an enormous, almost horizontal thrust fault (an overthrust). But soon, he began to doubt his courageous hypothesis, even though his British colleague Sir Roderick Impey Murchison found his idea both logical and convincing. Instead, he began to construct a double-fold structure resembling a tobacco-pouch that presumably formed because the globe is shrinking, a generally accepted assumption at that time. His student and successor Albert Heim (1849–1937) adopted this interpretation and defended it with *verve* against all criticism. In 1891, he finally

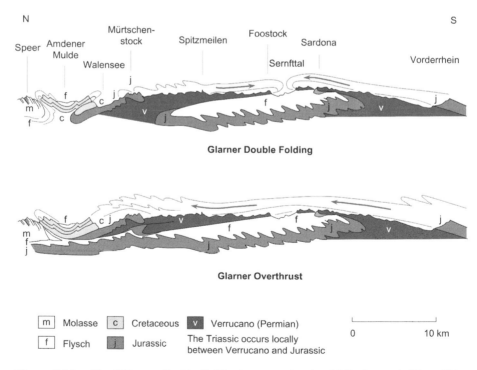

Figure 2.14 *The 'Glarner Double-Folding' as seen by Arnold Escher and Albert Heim 1870–1902 (top) and the Glarner Overthrust as seen by Marcel Bertrand 1883 and Eduard Suess 1892 (after Heim 1921)*

published his masters' interpretation in his book *Geologie der Hochalpen zwischen Reuss und Rhein* (*Geology of the Higher Alps between the Rivers Reuss and Rhine*). At that time, "Glarner Double-Folding" (Figure 2.14) was already a geological dogma.

But Marcel Bertrand, teaching geology at the École des Mines in Paris, had already drawn a profile in 1884 explaining the geological situation with a veritable overthrust even though he had never visited the terrain. August Rothpletz from Munich and Eduard Suess from Vienna consented to Bertrand's interpretation. Eventually, Hans Schardt from Neuchâtel proved that the Swiss Préalpes had been thrust from central Switzerland and Maurice Lugeon, a renowned geologist from the Canton Vaud, agreed with him. Albert Heim had to admit that Bertrand was right: the overthust theory became generally accepted (Figures 2.14 and 2.15) and has served ever since to explain complex tectonic problems.

This example demonstrates how demanding the reconstruction of geological profiles is. Mountain building processes, consecutive folding and faulting often obscure the original setting and make any unambiguous interpretation difficult, sometimes even impossible. In the following, the basic tectonic elements are introduced.

A *fold* is a geological stratum that has been bended due to tectonic forces. Upward folds are called anticlines and downward folds synclines. The axial plane consists of fold axes and divides the fold into two limbs. Folds may be symmetrical, asymmetrical, inclined and overturned with horizontal or plunging fold axes, as illustrated in Figures 2.16 and 2.17.

Figure 2.15 *The Glarner Overthrust as seen from Elm (Canton Glarus); the older Verrucano lies over the younger Flysch*

31

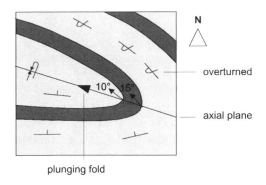

plunging fold

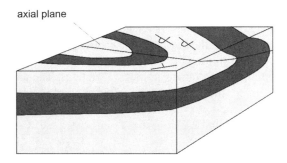

Figure 2.16 *A SSW-inclined fold*

In addition to folding, strata are fractured into *discontinuities* that can further be classed as:

- faults
- joints
- cleavage and schistosity
- unconformities and bedding planes.

Faults are discontinuities along which a displacement has taken place, as illustrated in Figure 2.18. When rocks above the fault plane move down in relation to the rocks above, it is known as a normal fault. If rocks above the fault plane move up in relation to the rocks below, it is referred to as a reverse fault. Normal faults are associated with extension, whereas reverse faults are related to shortening. With respect to their relative movement, they are also referred to as:

- dip-slip faults, when one formation moves up or down along the fault plane
- strike-slip faults, when one formation moves laterally, and
- oblique-slip faults, when both relative movements are combined

as depicted in Figure 2.19. Folding and faulting usually interacts with each other to form complex patterns (Figure 2.20).

32

Figure 2.17   *Folded rock in the Allgäu, Germany*

(a)   (b)

Figure 2.18   *(a) A fault in Andalusian sediments, Spain and (b) a fault in a Naukluft rock formation, Namibia (section about 4 m²)*

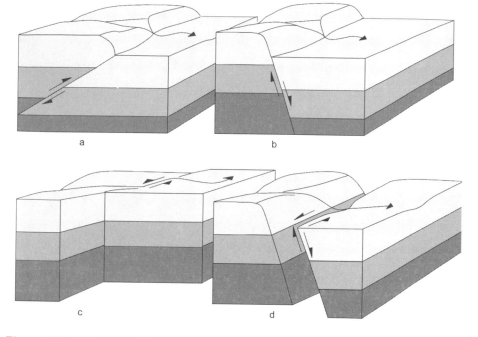

a   b

c   d

Figure 2.19 *(a) Reverse fault, (b) normal fault, (c) strike-slip fault and (d) oblique-slip fault*

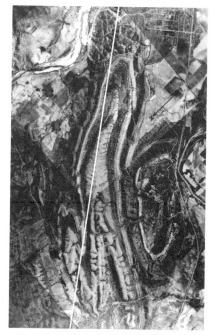

Figure 2.20 *Aerial photo of a plunging fold disturbed by a strike-slip fault as indicated with the white line*

In tension zones, which develop when continental plates drift apart for example, rock formations may be downfaulted along normal faults to form tectonic grabens. On the other hand, in zones of compression rock formations are upfaulted to form horsts. In certain cases one formation can override the other along almost horizontal thrust faults. In that case, a nappe of older rocks may cover younger rocks like a sheet with eroded windows revealing the younger stratum. Faults may be open and conduct large amounts of groundwater or be filled with broken rock; they may be plugged with weathered material or healed with minerals. Faults are always to be considered as potential planes of weakness that may cause stability problems and rapidly disperse contaminants from, for example, a leaking landfill. The direct vicinity of faults is often disturbed and fractured.

Joints are discontinuities along which no displacement has taken place, and can be the results of tectonic forces as with faults. A terminology to classify joints was introduced by Sander (1930) who proposed the abc-joint classification system (Figure 2.21) to annotate discontinuities in a fold structure: the b-axis is aligned parallel to the fold axis, the c-axis perpendicular to the bedding and the ab-plane parallel to the bedding. In addition, the hkl-system permits specification of diagonal joints (Figure 2.21): those producing an a-axis as a line of intersection are referred to as 0kl-joints, those with a b-axis are labelled h0l-joints and those producing a c-axis are called hk0-joints. As well as tectonic joints, other types are possible such as those caused by stress release (e.g. due to erosion of covering strata, melting of glaciers, etc.), cooling (e.g. when basalt cools off), shrinking (e.g. when clay dries out) or freezing and thawing (e.g. in permafrost soils), to name a few. Figure 2.22 depicts fractured and jointed outcrops.

Furthermore, *cleavage* and *schistosity* as found in deformed bedrock constitute special types of discontinuities from which the direction of tectonic forces can be deduced.

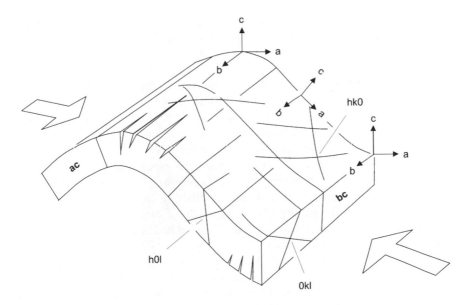

Figure 2.21 *Tectonic joints: abc-system and hkl-system*

(a)

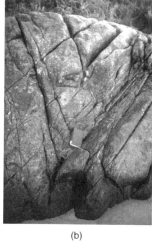

(b)

Figure 2.22 *(a) Fractured sedimentary rock at Chapman's Drive close to Cape Town, South Africa (section about 10 m²) and (b) jointed igneous rock in Queensland, Australia (section about 4 m²)*

Surfaces of erosion or non-deposition are marked by *unconformities*. The rocks below the unconformity are older than the rocks above. The unconformity itself represents a break in time, and information on what was eroded during the lacking record must be gained indirectly, for example from the eroded material that may have accumulated in sedimentary basins adjacent to the location of the unconformity.

Fluctuations in sedimentation rates and sediment types produce stratification. *Bedding planes*, exposing sedimentary structures such as ripples, separate different strata. These can represent a former beach, riverbed or sand dune. Bioturbations indicate the activity of former organisms. Typically, a stratum begins with coarse sediments that become finer towards the top (graded bedding), but many different sedimentation milieus can be observed.

Just as in bedrock, a variety of discontinuities can be found in uncompacted sediments and also soils. They constitute planes of weakness and preferred flow and transport paths. It is therefore important to identify and characterise all sets of discontinuities and to interpret their pattern with regard to possible implications for the planned project.

3 Functions

Ground serves us in a number of different ways, from being simply building ground to providing an environmental buffer system. It follows that ground is a precious good that has to be preserved and protected.

3.1 Functions of natural ground

Natural ground exposes a number of important functions that can be divided into three groups:

- ecological functions
- functions of exploitation
- functions of documentation.

These functions denote the fundamental importance of ground to our society. The ecological functions of ground have only been investigated in detail recently, although its economical value was recognised by the Scottish moral philosopher Adam Smith (1723–90) in 1775 with the publication of *An Inquiry into the Nature And Causes of the Wealth of Nations*, one of the first and most eminent introductions into modern economics. In his essay he stressed the importance of ground in the production of wealth. In fact, he reduced wealth to three vital production factors: capital, labour and *ground*.

3.2 Ecological functions

3.2.1 Soil as living environment

Soil is the living basis for plants and animals. Although life started in the oceans, plants and animals invaded the land some 400 Ma ago. Since soil provides many resources and nutrients, a broad variety of terrestrial ecosystems have emerged.

Soil organisms produce *humus* (Latin for moist, productive soil) from organic matter, which makes soil fertile. They also reduce organic matter to water, carbon dioxide and minerals from which plants profit. Larger soil organisms such as worms, ants or groundhogs introduce pores that may comprise 35–69 % of the volume of the topsoil, which is watered and aerated via these pores. Soil water acts as a transporting agent for nutrients and, together with soil air, enables the complex biochemical processes essential for the survival of all soil organisms. The surface area of all the soil particles in a 30-cm cube would cover an area up to 10 km^2. One handful of topsoil accommodates a larger number of soil organisms than the number of people living on the planet. Today, only 10 % of them have been identified and taxonomically recorded.

The *microflora* (0.6–50 μm) constitutes the biggest part of the biomass of the upper soil zone. In fact, bacteria, fungi and algae make up 80 % of the organisms living within the topsoil. Likewise, they are responsible for 90 % of soil metabolism (see Table 3.1). In this respect, the smallest by size drive most biochemical transformation processes whereas larger soil organisms control and enhance them. Bacteria, fungi and algae interact and cooperate with protozoa, insects, worms and large animals such as moles and snakes. They form a multitude of symbioses, the most important involving the roots of vascular plants and mycelium as a fungal partner. The mycorrhizae-symbiosis allows plants to absorb nutrients within the *rhizosphere*, the soil space dominated by the roots of the plants. Consequently, terrestrial plants could not survive without soil organisms. Moreover, 95 % of all insects spend part of their life within the soil. All terrestrial animals take advantage of the resources soil offers, just as we do.

These functions and capacities are not obvious from a soil profile (Figure 3.1). It is perhaps because of this that we are so reluctant to recognise the value of the resource soil. Since the appearance of *homo sapiens*, the natural balance between soil and the living environment has weakened. Humans exploit the ground to an extent previously unseen. In many cases, this exploitation is not motivated by the satisfaction of basic needs, but instead profit-driven. With an increasing population, the amount of soil available to satisfy everybody's demands diminishes. The ecological footprints left by those living in high-income countries exacerbate the situation and destabilise the global society that is already living in conflict with nature.

Table 3.1 *Soil as living environment (after Fründ and Meuser 2000)*

	Microflora	Microfauna	Mesoflora	Macrofauna	Roots
Diameter	0.6–50 μm	20–200 μm	0.2–2 mm	2–20 mm	> 10 μm
Species	Bacteria, fungi, algea	Protozoa nematoda, etc.	Enchytrae, mites, springtails, etc.	Earthworms, woodlouses, snails, bugs, flies, spiders, etc.	Roots
Biomass	> 80 %	< 20 %	< 20 %	< 20 %	< 20 %
Metabolism	> 90 %	< 10 %	< 10 %	< 10 %	< 10 %
Mobility	mm–cm	mm–cm	cm–dm	> m	> m
Regeneration	Hours	Days	Weeks to months	Months to years	Weeks to years
Ecological role	Biochemical transformation	Control and amplification	Control and amplification	Control and amplification	Control and amplification

Figure 3.1 *Soil profile (Hierapolis, Turkey)*

3.2.2 Soil as a recharge and buffer system

As well as functioning as the basis of living for humans, animals and plants, the ground also exchanges and recycles fluids and gases. For example, groundwater resources are recharged by the infiltration of rainwater (Figure 3.2). Once water begins to percolate into the ground, it will pass through a zone that is still aerated. At this stage the soil water is referred to vadose water. As soon as all soil pores are filled with water, the soil water is referred to as groundwater. The upper boundary of the groundwater is called the groundwater table. This recharge process depends largely on the properties of the soil, especially porosity and permeability. Certain soil types like podzols, cambisols or arenosols allow high infiltration rates whereas alisols, lixisols or plinthosols tend to become waterlogged. The tropical vertisol is a typical example of soil exposing a high initial infiltration rate due to cracks developed during the dry season. These cracks rapidly close when the rainy season begins, however, since certain clay particles within the A- and B-horizon increase in volume upon contact with percolating water (see Chapter 4).

Rainwater that cannot percolate into the soil leaves the catchment area as overland flow and accumulates in rivers and streams (Figure 3.3). Surface runoff increases with increasing density and clay content of the soil and with increasing inclination of the surface. It decreases, however, as soon as vegetation establishes. Infiltrating rainwater is consumed within the rhizosphere and later transpired through the leaves. In addition, rainwater that is temporarily caught on the leaves may evaporate before it can actually enter the soil. These combined processes are referred to as evapotranspiration. The collection of precipitation by plants is called

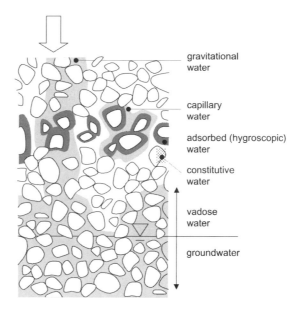

Figure 3.2 *The infiltration of water into the ground in the process of groundwater recharging*

interception, a process that is complemented with the interim storage of rainwater on decaying leaves that litter the ground (Figure 3.4). During the winter season, precipitation may be prevented from infiltrating the soil due to freezing conditions or snow covering.

Another example of the recharge capacity of soils is the carbon cycle. Plants take up carbon dioxide through photosynthesis and store it in the form of organic carbon. Carbon is also stored within the soil organisms as well as in plant residues and decaying organic material. Soil micro-organisms break down the organic carbon into

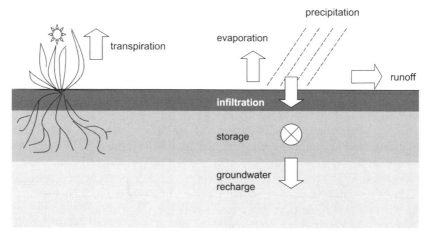

Figure 3.3 *Infiltration and recharge of groundwater*

Figure 3.4 *The eucalyptus tree wastes leaves and even whole branches to protect its own water reservoir from evaporation. When planted for pulp or firewood as in Spain or Brazil, the dense cover of litter intercepts infiltration, breaking the groundwater recharge cycle.*

water and minerals while releasing carbon dioxide. If this process is balanced, there is no net release of carbon dioxide into the atmosphere. If more carbon is consumed than released, soil acts as a carbon sink. The degradation of the world's climate indicates, however, that the opposite is the case: every year about 2.3 Gton of carbon is sequestered in the terrestrial system and about the same amount is stored in the oceans. In contrast, 6.3 Gton are released every year as emissions from fossil fuel use and 1.6 Gton are released due to the change of land use, notably the reduction of the rain forest. Consequently, there is a yearly accumulation of 3.3 Gton in the atmosphere, which is considered to be one of the main reasons for the greenhouse effect (IPCC 2000). The rise of the mean temperature may have side effects that we so far cannot predict. For example, with increasing temperatures the cryosols of the permafrost regions may thaw, prompting the formerly frozen organic material to be broken down and mineralised by micro-organisms while large amounts of carbon dioxide and methane are released.

Soil also buffers contaminants which originate from industrial production, waste storage, accidents etc. A broad variety of toxic substances are released every day including organic contaminants like solvents and fuels as well as inorganic substances like heavy metals and asbestos. Soil takes up these contaminants, retains them and partly neutralises them in three different ways:

- mechanically
- chemically and
- biologically.

The covering layer of organic matter and the upper soil horizons mechanically retain considerable amounts of sediments and dispersed matter. The porous soil matrix filters the percolating runoff, thus keeping many contaminants from entering the groundwater. Excessive infiltration of particles may lead to colmation i.e. clogging or blocking of the pores, causing a decrease in permeability. Suffusion (underground erosion) may subsequently restore the original permeability while releasing the trapped particles. Fine-grained soils have a high retardation potential while coarse-grained soils are less likely to retain contaminants.

Both organic and inorganic contaminants may adhere to the soil particles, especially the surface of clay minerals, humus and metal oxides. This process is called adsorption. The equilibrium between the concentration of contaminants within the groundwater c (mg l^{-1}) and the mass of contaminants adsorbed to the soil matrix c_S (mg kg^{-1}) depends on the distribution coefficient or partition coefficient K (ml g^{-1}) and the exponent ϕ (dinemsionless) (Freundlich 1907):

$$c_S = K c^{\phi}$$

This equilibrium is governed by ambient temperature and is therefore referred to as the *Freundlich isotherm*. It describes the adsorption equilibrium at a constant temperature. The mass adsorbed usually decreases with increasing temperature. K and ϕ are listed in standard references for various contaminants (e.g. Schwarzenbach *et al.* 2002). $\phi > 1$ implies that considerable amounts of contaminants may be adsorbed to the soil matrix. $\phi < 1$ refers to a weak adsorption capacity (Figure 3.5).

As the adsorption of organic contaminants is a function of the organic carbon content of the soil, K can be related to the fraction of organic carbon in the soil f_{oc} and the organic carbon partitioning coefficient of the contaminant K_{oc} (ml g^{-1}):

$$K = K_{oc} f_{oc}$$

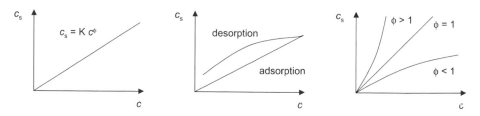

Figure 3.5 *(a) Linear adsorption, (b) adsorption and desorption hysteresis and (c) non-linear adsorption with Freundlich exponents*

If K_{oc} is not known, it can be calculated from the octanol-water partition coefficient K_{ow}:

$$\log K_{oc} = a \log K_{ow} + b$$

where examples of constants a and b can be seen in Table 3.2. This allows a straightforward assessment of the amount of organic contaminants adsorbed to the soil.

Adsorption of contaminants to the soil matrix (LaGrega *et al.* 2001)

Groundwater near a leaking underground tank contains 0.5 mg l^{-1} benzene. What is the expected concentration of benzene adsorbed to the silty soil that contains 2 % of organic matter?

Solution

For benzene, K_{oc} is listed as 83 ml g^{-1}. Thus, it follows that

$$K = K_{oc}f_{oc} = 83 \times 0.02 = 1.66\text{ml g}^{-1}$$

With the Freundlich exponent ϕ assumed to be 1.0, the adsorbed concentration follows from

$$c_S = K\,c = 1.66 \times 0.5 = 0.83\,\text{mg kg}^{-1}$$

that is, 0.83 mg of benzene will be adsorbed to 1 kg of soil.

As well as the capacity of soils to adsorb contaminants, soils may also buffer acids, especially those soils stemming from calcareous bedrock (limestone, dolomite) and those accumulating secondary calcium carbonate (calcisols). A common practice to reduce the pH in acidic soil is the application of lime, either at the surface or by mixing it with the upper soil layer. Liming campaigns can be combined with measures to improve the fertility of the soil in order to enhance environmental conditions for plants

Table 3.2 *Coefficients to determine K_{oc} with K_{ow} (after Schwarzenbach et al. 2002, LaGrega et al. 2001)*

	a	b
aromatic hydrocarbons	1.01	−0.72
chlorinated hydrocarbons	0.88	−0.27
pesticides	0.54	1.38
chloro-s-triazine	0.37	1.15
chlorphenols	0.81	−0.25

to establish. If a vegetation cover is re-established on degraded acidified terrain, erosion is reduced and soil conditions are gradual improved (Chapter 9).

Soil organisms like microbes and fungi are capable of biodegrading organic contaminants and of mineralising them into inorganic residues, carbon dioxide and water. The process of biodegradation can be stimulated by adding nutrients to the soil, a technique referred to as bioremediation. Induced biodegradation processes are considered to be a well-established low-cost approach to remediate sites contaminated with organic pollutants (Chapter 9).

Mechanical, chemical and biological mechanisms of filtering, buffering and neutralisation may take place at the same time and may interact with each other. Consequently, it may appear difficult to predict the capability of soil to retain and eliminate contaminants under field conditions and empirical approaches are usually applied to judge the buffer capacity of a certain soil profile.

As soon as the retention, neutralisation and biodegradation capacities are exhausted, pollutants may pass the upper soil layer and migrate towards the groundwater table. The various types of contaminants, their origin and migration pattern will be discussed in more detail in Chapter 6.

3.3 Exploitation of ground

3.3.1 Agriculture

According to the European Environmental Agency, soils are the basis of 90 % of all human food, livestock feed and fibre. About half of the EU's land area is exploited by agriculture. The fertility of soil depends on physical, chemical and biological characteristics. The physical characteristics include:

- structure and
- texture.

The properties of the soil affected by these characteristics are listed in Table 3.3. Physical characteristics also determine the infiltration rate and the ability of the roots to penetrate the ground and access nutrients and water. The chemical characteristics include:

- availability of elements essential for plant growth such as carbon, hydrogen, oxygen, nitrogen, phosphorus, calcium, magnesium, sulphur, potassium, chlorine, iron, manganese, copper, zinc, molybdenum, boron and others
- salinity, which affects the mobility of water within the soil matrix

Table 3.3 *Soil properties affected by soil structure and texture (after Raven et al. 1998)*

	Aeration	Drainage	Water-holding capacity	Nutrient-holding capacity	Workability (tillage)
Sandy soil	Excellent	Excellent	Low	Low	Easy
Loam	Good	Good	Medium	Medium	Moderate
Clay soil	Poor	Poor	High	High	Difficult

- pH-value, which affects the solubility of minerals and the availability of nutrients to be absorbed by plants. A pH between 6.5 and 7.5 is considered favourable for the fertility of soils.

The biological characteristics include:

- amount, types and diversity of soil organisms
- amount, type and distribution of organic matter that soil organisms feed on and break down
- symbioses that develop within the soil, especially the microbial processes within the rhizosphere.

The availability of water is a prerequisite for all physical, chemical and biological processes. Water is stored temporarily within the rhizosphere where it becomes available for plants. The ability of soil to retain moisture is called field capacity and is defined as the capacity to store moisture after complete wetting and after free drainage becomes negligible (Or and Wraith 2000). The usable field capacity, on the other hand, refers to the amount of water available to plants (see Table 3.4). These parameters are not correlated (see Figure 3.6), since soil water in finer pores (dead water) is subject to increased water tension and consequently cannot be consumed by plants. Clay, for example, has a high field capacity but a low usable (i.e. plant-available) field capacity. The usable field capacity thus depends on the size and the total volume of the pores, the distribution of pores and grains within the soil matrix and the amount of humus produced. For example, chernozems and luvisols typically have a high usable field capacity whereas arenosols and podzols are reduced in their capacity to retain water.

When the transpiration rate of a plant exceeds the amount of water it can take up from the root zone, the wilting point is reached and the plant will fail to recover its turgidity. Wilting point and usable field capacity control the amount of water available for plants. Together with the availability of nutrients, they constitute the key parameters for the fertility of a soil. If a soil is not farmed, the nutrients extracted by plants and organisms are cycled back into the soil. If the soil is farmed, the nutrient cycle is broken and fertilisers are needed to keep the soil productive. Figure 3.7, for example, depicts the number of years until the mineral reserves of farmed soils in France are exhausted if no minerals are added. (The upper and lower limits are based on mean mineral reserves and annual extraction rates due to harvesting, according to data presented in Barbault 2003). The excessive application of fertiliser, however, degrades the soil just as excessive farming does, as will be shown later.

Table 3.4 *Water storage capacity and water availability for plants (after AGB 2005)*

	Maximal (% of volume)	*Usable* (% of volume)
Sand	15–31	11–24
Loam	32–38	10–16
Silt	34–39	20–26
Clay	36–42	9–15

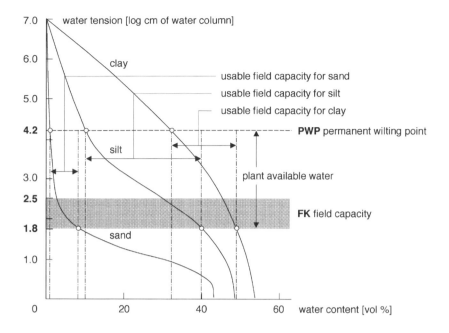

Figure 3.6 *Field capacity and usable field capacity for different soil types*

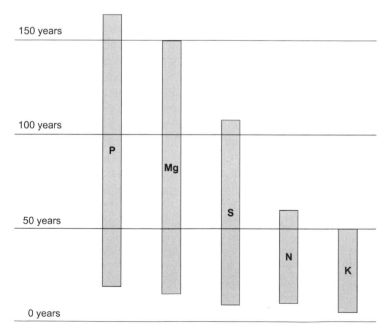

Figure 3.7 *Number of years until mineral reserves of farmed soils in France become exhausted, if none are added*

3.3.2 Mining

"If we can't harvest it, we'll mine it!" reads the typical bumper sticker on the cars of those working for the mining industry. A great variety of resources may be extracted by mining, including

- topsoil as re-cultivation substrate
- sand and gravel as aggregates
- building and ornament stones including sandstone, marble and granite
- roof cover such as slate
- limestone, potash and rock salt for industrial production
- coal, oil and gas as energy sources
- ore to smelter and refine.

Depending on the type of resource to be extracted, different mining methods are distinguished:

- surface mining or open pit mining where the overburden (the soil and low grade material) is removed to obtain the resource
- subsurface mining or deep mining where shafts are sunk and galleries are mined to reach the resource
- pumping of liquids and gases.

Mining began in prehistoric times, when resources were either at or close to the surface. From the Brecklands of East Anglia, England, to Arnhofen in Bavaria, Germany, Neolithic people dug for flint in bell-shaped pits about 10 m deep. They required up to 200 hrs to sink a pit, where they extracted (using horns and antlers) several kilos of flint stone that were then split into about a hundred blades for knifes, sickles and weapons. Once a pit had been exploited, another one was dug right next to it and vast pit fields developed that are still detectable today. With an increasing population, the need for resources increased and mining methods became more refined. The scientific basis of modern mining was set by Georgius Agricola (actually Georg Bauer 1494–1555) with his book *De Re Metallica Libri XII*. Soon, sophisticated mining technologies emerged, especially after the invention of the steam machine that allowed the miners to control the groundwater and thus reach deeper into the ground.

Today, we distinguish between room-and-pillar mining (a pattern of galleries that cross each other leaving pillars to support the roof of the mine) and longwall mining (the mineral is sliced off along a longwall that connects two galleries). There are also special mining techniques, including open stope mining to extract minerals along geological veins and solution mining to extract rock salt (Chapter 7).

In 1991, the Worldwatch Paper 109 *Mining the Earth* stated that more raw material was mined than all rivers on this planet would erode into the seas. The exploitation and refinement of our resources account for one-tenth of the world's energy consumption and causes a stream of waste much larger than any other manmade waste stream. This is of course accompanied by an enormous depletion and contamination of land as Figures 3.8 and 3.9 amply demonstrate.

Figure 3.8 *Mining expanded in the German Ruhr District, consuming fertile ground (after Block 1928)*

Figure 3.9 *Devastated landscape due to mining in Rio Tinto, Spain*

3.3.3 Settlement and urbanisation

With increasing population, the demand for land to develop settlements has been increasing. Cities with their ever-growing suburbs and satellite towns occupy large areas.

The 21st century will be dominated by urbanisation, defined as an increase in urban population which is greater than the increase in total population. The demographic trends indicate that urban population will double to some 5 billion from 2000–25. In 2025, two-thirds of the world's population will be living in cities. Figure 3.10 illustrates the percentage of the world's populations living in cities. In 2015, it is predicted that there will be around 360 cities of more than 1 million, about 150 of which are located in Asia. As Hall and Pfeiffer (2000) state in their World Report for the Urban Future 21, 27 megacities of more than 10 million inhabitants will be established, of which 18 will be located in Asia (Figure 3.11). Figure 3.12 depicts the apparently never-ending skyline of Sao Paulo in Brazil, which is predicted to experience an increase in population to 21 million by the year 2015.

Hall and Pfeiffer distinguish three types of cities:

* *Cities characterised by spontaneous, excessive growth.* Examples are many cities in Africa south of the Sahara, in India, in the Middle East and in the poorer regions of Latin America. These cities grow rapidly due to an in-migration from rural areas and high birth rates. Their economic potential, however, is inhibited by mismanagement and corruption, facilitating the development of large unofficial settlements lacking in economic stability and environmental protection.

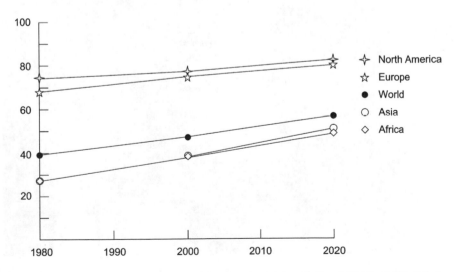

Figure 3.10 *Percentage of populations living in cities (after Hall and Pfeiffer 2000, Schubert 1999)*

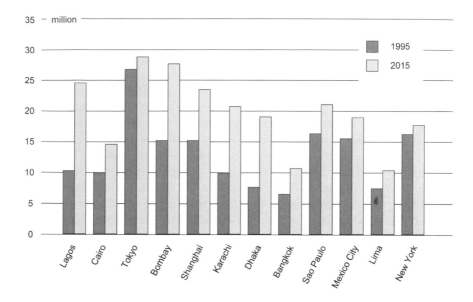

Figure 3.11 *In 2015, there will be 360 cities of more than 1 million inhabitants and 27 of more than 10 million (after UN population statistics)*

Figure 3.12 *Viewed from one of the highest buildings of Sao Paulo, the 168 m tall Edificio Italia, the city extends to the horizon*

- *Cities with a dynamic growth pattern.* Examples are cities of middle-income countries in Eastern Africa, Latin America, the Caribbeans and the Near- and Middle East. The population does not grow as dramatically as in the cities of the first category, which releases economic and ecologic pressure and leaves space for economic growth.
- The *mature cities* of Northern America, Europe, Asia and Australia. These cities suffer from an increasingly aging population, a trend that is globally observed but that principally inflicts upon the economies of mature cities. Nevertheless, many of these cities are comparatively wealthy and still have the capacity to renovate their infrastructure and mitigate the adverse effects of maturation.

It will be these three types of cities that will govern life in the 21st century. They will change economic growth patterns and will impose on our environment. They may degenerate into sterile megalopolitan corridors (Hall 1999) or improve to lively and wealthy urban spaces in line with the ideas of a sustainable city renaissance.

While the city council of New Delhi has given up attempting to control the expansion of the city in an organised way, and has retreated to observe it by means of satellite photographs, the urban planners of the industrialised nations follow a policy of suburbanisation that threatens the remnants of nature left around the old metropoles. In Germany, for instance, urban sprawl still consumes about 100 ha day^{-1} (BBR 2005). This is equivalent to 10 m^2 s^{-1} or 340 km^2 yr^{-1}, an area equal to that occupied by the city of Munich. Although the population in Germany has been rising only 20 % during the last 50 years, land consumption has doubled. In 1991, the United States became the first nation where the total suburban population exceeded both city population and rural population. While new streets are constructed and new supply lines are laid to satisfy the demands of the 'grease belts' of the cities, half of the space consumed by the expansion of the poor cities in low-income countries will be occupied by informal settlements (UN-Habitat 2003). Figure 3.13 depicts the predicted rise in urban sprawl for some of the world's most populated cities.

We also have to note that our society is aging, not only in the industrialised nations but also in countries in transition and low-income countries. By 2030, about 16 % of the world population will be older than 60. In North America, one out of four will be older than 60, while in Europe almost every third person will have exceeded that age. This trend can also be observed in Asia and Latin America, where by 2030 almost every fifth person will be older than 60, while in Africa this level will be reached by 2050 (Höhn 1996). There are many reasons for this development including improving health standards and care options for senior citizens. Directly coupled with this development is a growing demand for housing space. Older people tend to stay in the dwellings they have occupied in their previous family phase (Schubert 1999). Furthermore, children tend to leave their parents' home earlier to study, work and live individually in their own apartment. Young families tend to have fewer children. Consequently the number of members per household decreases, while the number of households increases continuously. This is especially valid for urban households, accelerating the land demand of our cities. An increasing demand for appropriate infrastructure, to cope with the changes in the demographic pattern, is also evident.

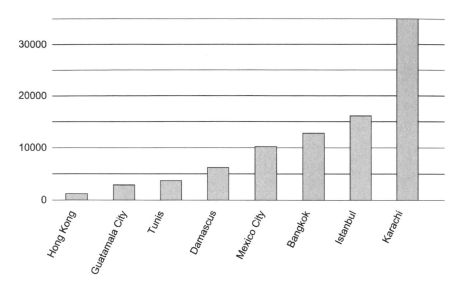

Figure 3.13 *Infrastructure demand in hectares from 1990–2015 (Compendium of human settlement statistics, United Nations New York 1995)*

All these trends and tendencies result in an increasing demand for urban land all over the world. It has already been stated in 1995 by the United Nations that by 2015, Mexico City will need another 10 000 ha, Bangkok another 13 000 ha and Karachi some 35 000 ha (Figure 3.13). This trend was confirmed during the 2nd UN Conference of Human Settlement in Istanbul in 1996 (Habitat II), and was again at the top of the agenda at the World Urban Forum 2004 in Barcelona.

On the other hand, according to recent studies, many cities demonstrate a tendency to shrink, from California to Japan and from Finland to the Cape of Good Hope. The reasons for city shrinking are manifold including de-industrialisation and post-socialism. It has also been observed that the fertility rate has been dropping significantly, a phenomenon which is again linked to urbanisation and which has been confirmed for both high- and low-income countries. The global mean fertility has been dropping from 5.4 in 1970 to 2.9 in 2000. In order to stabilise the population of a country against war, famine, epidemic and disaster, a birth rate of 2.1 children per woman is needed. However, the population is known to be declining in certain areas, as depicted in Figure 3.14. In Europe, examples of shrinking cities can be found especially at past industrial belts and in regions that are economically declining (Oswalt 2005, Oswalt and Rieniets 2006). In some East German cities, for example, entire neighbourhoods are dismantled after the population has left to seek work in other parts of the country (Figure 3.15).

We conclude that we are presently experiencing a period of urbanisation with first indications of urban shrinking. In times of global markets and local bottlenecks, cities change constantly and rapidly, much faster than they used to do even in the boom-days of industrialisation. Changing cities have a strong impact on both economy and ecology. The metamorphosis of urban spaces usually degrades the environment and

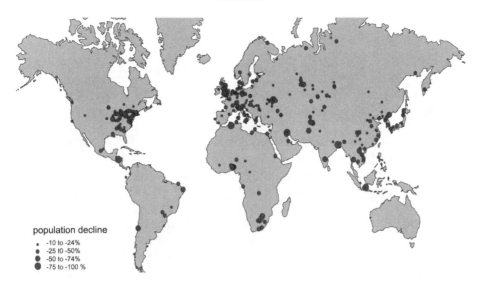

population decline
- -10 to -24%
- -25 t0 -50%
- -50 to -74%
- -75 to -100 %

Figure 3.14 *Shrinking cities (after Oswalt 2006)*

Figure 3.15 *Dismantling of apartment blocks in Wismar, Germany, a city with a declining population (photograph courtesy of Ariane Ruff)*

exhausts the resources. However, changing cities also offer chances, not only to investors but also to new forms of land and resource management.

Shrinking phenomena in low-income countries

City shrinking may also be associated with environmental degradation and the absence of sufficient resources, especially water and energy. Although access to water has somewhat increased over the last decade with only 15 % of the developing world left unserved, sanitary conditions have not improved at all. Today, more than half of the world population has no access to even the most basic sanitary infrastructure. In countries such as Guinea-Bissau, only 50 % of the population have access to safe water. There is a child mortality rate of about 20 % i.e. one child out of five dies before the age of five. Every year, approximately 2 million people die from diarrhoeal diseases and many more are suffering at any one time from one or more of the six main diseases associated with hygiene, sanitation and water supply, including diarrhoea, ascaris, dracunculiasis, hookworm, schistosomiasis and trachoma (Hueb 2000). Strategies on the sanitation for high risk communities have been introduced by the World Health Organisation WHO (Genske *et al.* 2000).

The global AIDS crisis underlines the gravity of the problem. Today, about 40 million people are infected with the deadly virus. The pandemic hits the poorest regions in the world and is spreading at a menacing pace. In the south of Africa, the number of orphans has increased to 12 million. In Botswana, the life expectancy has fallen to 40 years. In Zambia, the number of teachers dying from AIDS every year is twice as large as the number of teachers that the national universities can generate. A severe famine threatens Malawi, since workers are missing the crop harvest (Grill 2004). After devastating the south of Africa, AIDS is now raging Russia, China and India. In 2005, 8500 people died every day. This is equivalent to the World Trade Centre being destroyed three times every day.

3.3.4 Waste dumping

Another example of the exploitation of land is the dumping of waste. Ever since mankind emerged, waste has been produced. In mediaeval times, waste was simply left in the streets, often thrown out of the windows. With increasing populations, public health degraded and epidemics broke out, prompting early efforts to systematically collect and dispose of waste. In 1184, the French King Phillippe II was nauseated by the disgusting odours of Lutetia, the old name for Paris meaning 'City of Mud', and ordered streets to be paved and cleaned. He did not meet with much success, since only two streets were finally paved and cleaning was quickly abandoned (Lieberherr-Gardiol 1997). Even after the Black Death devastated the city in 1348, the newly introduced laws regarding street cleaning were not followed. 'Cleaner pigs' eating the edible residues were roaming the streets of Paris and other metropoles. In fact, they were still found in New York City until the 19th century.

It was not until 1883 that the Prefect of Paris, Monsieur Poubelle, ordered his citizens to dispose of their waste in 40–120-litre steel or metal-lined wooden containers to protect the city from vermin and epidemics. He also obliged his citizens to sort their waste into glass, paper and compostable waste. Although it took the Parisians until 1940 to finally accept the system, Monsieur Poubelle must be considered as one of the first to dispose of waste in an organised way or to promote the idea of recycling.

In those days, the production of waste was still comparatively low, since organic wastes were used as fertilisers or consumed by household animals, glass bottles were reused and metals, cloths and other items were collected by waste pickers to be reworked for reselling at the rag-and-bone trade. However, with the beginning of industrialisation waste streams multiplied, and large amounts of noxious wastes were dumped in derelict quarries or heaped right next to the company. This was accompanied by an enormous consumption of land.

At the beginning of the new millennium, Europeans produce some 400 kg of household waste annually. US citizens generate almost twice that amount. Even in Switzerland, a country renown for its environmental awareness, an urban family of four produces more than 100 ton of household waste during their lifetime, requiring more than 200 m³ of landfill space (Figure 3.16). Today, it has become common practice to sort and recycle waste either to extract secondary resources such as paper and glass, or to produce energy by incineration. Biodegradable green waste is converted into biogas and compost. The potential space consumption for landfills has thus been greatly reduced. Still, the dumping sites of the past scatter our landscape and will continue to cause problems for the generations to come.

A large variety of different waste types can be distinguished and many different ways to groups these wastes have been developed. One may distinguish between the

- physical state
- origin

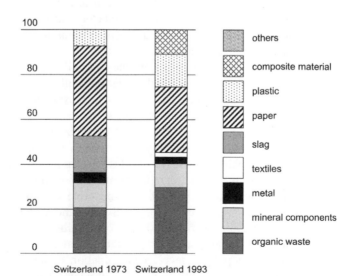

Switzerland 1973   Switzerland 1993

Figure 3.16 *How composition of household waste in Switzerland has changed over 30 years (after BFS 2002)*

55

- level of danger
- potential to be recycled, downcycled or transformed into secondary resources.

Wastes may originate from

- households producing food and yard wastes, paper, plastic, glass, metals, ceramics, wood, textiles, appliances, etc. (municipal solid waste, or MSW)
- businesses and industries producing wastes specific to their production lines
- construction sites producing debris and rubble from dismantled buildings as well as polluted soil to eliminate contamination hot spots
- mining sites producing waste rock and effluents from ore processing.

Hazardous wastes have been of concern because of their potential to degrade our environment and threaten our health. They are grouped as

- toxic wastes dangerous to the environment
- corrosive wastes having a pH < 2 or > 12.5
- reactive wastes provoking chemical reactions, fires and explosions
- ignitable wastes that can cause fires under standard temperature and pressure conditions.

With regard to stocking wastes in landfills, three groups of wastes are distinguished:

- inert or inactive wastes
- wastes that leach hazardous substances
- active wastes producing hazardous substances.

Only the first group of wastes can be stored easily. In order to dump active wastes and wastes that leach hazardous substances, landfills have to be sealed and covered to avoid contact with people and goods to be protected, such as the groundwater. In addition, complex gas and leachate management systems need to be installed that require maintenance for many years.

Land consumption and soil destruction due to landfilling is certainly a crucial aspect of human-induced environmental degradation. With increasing population, the amount of land occupied by landfills has also been increasing. In former times, wastes were stored just outside the cities. Since most of the cities have increased in size, many landfills have finally become part of the city, obstructing their development while imposing environmental risks. Only recently, measures of waste reduction and recycling have been established in national rules and regulations. In certain cases, it has even been decided to excavate derelict landfills to eliminate the contamination source and create new building ground. A remarkable example is the Hazardous Waste Landfill Kölliken in the Canton Aargau, Switzerland, which was established in May 1978 to receive more than 250 000 m³ of hazardous wastes from different chemical industries. Although placed in permeable Quaternary formations, no base liner was installed, prompting a severe contamination of the adjacent aquifer. In addition, emissions of toxic landfill gases have caused expensive remediation works. The intended excavation of this landfill calls for a budget exceeding 500 million euros (Genske 2003).

Dumping wastes by Swiss chemical industries (Forter 2000)

Until the Second World War, the Rhine River served as a dustbin for the City of Basel's chemical industries. Before 1860, production refuse—a good proportion of which could be considered as hazardous wastes—were stocked at the banks of the Rhine behind the factories. Then, for about ten years, it was dumped through a hole in the middle bridge directly into the Rhine. Later the waste was ground, and washed with sewage into the Rhine or dumped into the river by special waste ferries.

The dry summers of the late 1940s, the increasing water demand of the city and the booming chemical industry of Basel finally brought waste managers to re-think their disposal practice. They were forced to halt their waste dumping routine and began to search for land-based disposal sites. Ciba Industry started to export their chemical wastes to the neighbouring city of Weil am Rhein in Germany. However, German environmental authorities soon banned this waste trafficking, concerned that groundwater resources would be spoiled for good. Ciba consequently had to divert its waste to the Swiss City of Muttenz 5 km southeast of Basel. Ciba and Geigy ignored the fact that the Muttenz pit borders the aquifer utilised by the City of Basel as a water reservoir. In 1955, the Swiss Water Protection Act was introduced, forcing Sandoz Industries, still following the convenient Rhine dumping practice, to haul its chemical waste to a French gravel pit at St Louis, just across the border. When, in 1957, serious contamination of the surroundings of the Muttenz pit became obvious to everybody, Ciba and Geigy followed the example of Sandoz and screened neighbouring French and German communities for handy dumping sites. They found them in the Hagenthal-le-Bas, Neuwiller, Hirschacker, Grenzach.

It took until 1960 for the German and French authorities to put an end to the illegal dumping business at their Swiss frontiers. Basel's chemical industries were left sitting on their waste, with neither Swiss nor foreign communities willing to take it. A dumping site further away was searched for and finally found some 40 km west of Basel in Bonfol, a small and poor village in the Swiss Canton of Jura, close to the French border. The old clay pit was quickly filled with chemical waste of which no record was kept. Six years later, the citizens of French Pfetterhouse complained of toxic clouds coming from the landfill. Ciba officially denied the existence of such clouds, while its own experts stated in internal reports that the contamination of the surroundings violates all civil regulations and environmental laws. Nevertheless, the Basel newspaper *Nationalzeitung* declared Bonfol an "exceptionally well-designed landfill".

3.4 Ground as archive

The function of the ground as an archive has been recognised by geologists, pedologists and archaeologists. The ground provides records of

- natural history and
- cultural history.

As a record of natural history, it documents the type and the structure of the parent material from which the soil was formed. With increasing maturation of the soil profile, the original material becomes more and more disguised by soil building processes. In certain cases, fossil soils or paleosoils (soils which have developed in geological times but have been covered again by sediments, on top of which new soils have formed) may be traced. As pedogenesis changes with temperature and humidity, past climate changes are recorded within the soil profile. Furthermore, former variations of the groundwater table can be read from the soil profile as well as many other natural processes that have influenced the soil forming processes.

Human activities have left many traces in the ground, from the earliest settlers of the stone ages to modern societies and their industrial heritage (for example, Figure 3.17). Ground records the relentless effort of our civilisation to extend its domain of influence at the expense of our natural environment. As an archive of cultural history, soil records for example early attempts to cultivate it and to improve its fertility (melioration). Plough traces of prehistoric settlers, as well as systematic patterns of boulders removed from the field and dumped along the fringes (as they would have been obstructing the ploughing) have been discovered in many parts of Europe. Humans drained and irrigated swamps and waterlogged fields. All these activities left marks in the ground that can be detected today.

In Europe, the human impact on the natural soil profile can be traced back to the Neolithicum. In the late Stone Age, farming was only carried out on isolated spots. Most of the early farmland was later abandoned, allowing the forest to recover.

Figure 3.17 *Vegetation patterns due to ground changes close to the ancient Pyramids of the Teotihuacán-Culture, Mexico, may hint at former structures so far undiscovered*

Figure 3.18 *A degraded soil profile near the former Bitterfeld industrial complex, Germany (photograph courtesy of Patrick Höhener)*

During the Bronze Age and the following Iron Age farming intensified, leading to initial and widespread soil erosion. The upper centimetres of farmed land were eroded and washed into the rivers. During the 5th and the 6th century, the climate deteriorated in Europe. It became colder and precipitation increased. This finally led to famine, wars and plagues. Tribes and nations started to migrate south, leaving behind their fields which were soon covered by forest again. During medieval times, land was again made arable and the erosion of the natural soil profile recommenced. The period between 1313 and 1350 appears to be of particular significance since heavy rainfall washed away about half of all the soil that was lost by erosion during the previous 1500 years (Bork 1989, Dotterweich *et al.* 2003). A special ploughing technique that left wide cambers between the furrows encouraged this rapid erosion. A similar period of heavy rain is reported from 1749 to 1800 (Chapter 5).

With the beginning of industrialisation the soil profile degraded further (for example, Figure 3.18). Toxic substances were released and migrated into the ground. In many places, soil was compacted or sealed. In order to create bearable building grounds, the upper soil horizons were removed and thus permanently destroyed. In urbanised areas, natural soil profiles are rare and the relicts left are difficult to protect.

4   Properties

In the foregoing chapters, the development of ground profiles has been explained and the functions of the ground presented. It has become clear how significant the ground is for the wealth and the cultural identity of our society. Soil has been interpreted as a system of three horizons with its upper horizon being in direct contact with the atmosphere, exposed to the agents of the climate and inhabited by complex symbioses of plants and animals.

From a more technical perspective, ground is defined as a deformable system consisting of solid matter, fluids and gases that can either be used as building ground or as building material. The properties of this system are discussed in this chapter.

4.1   Soil and rock

Underneath the active soil horizons (the solum with A-horizon and B-horizon), the parent material (C-horizon) is either represented by sediments or by bedrock. The physical properties of sediments divert considerably from those of bedrock. Sediments are characterised by layers of weathering products and rock fragments that may be as big as boulders or as small as clay minerals. Depending on the original material and the environment of sedimentation, these layers may be sorted or unsorted, homogenous or inhomogenous, consolidated or unconsolidated. Sediments are modelled as porous media whereas bedrock is modelled as a fractured medium. Bedrock is geologically compacted and fractured due to tectonic deformations and stresses. Only the intact portions between the fractures are actually referred to as rock whereas the ensemble is called rock mass. Rock fractures or discontinuities are due to tectonic stresses and deformations that the rock mass was exposed to and are thus systematic. However, as previously mentioned (Chapter 3), sediments may also show systematic patterns of discontinuities.

In the following, we will discriminate between *soil, rock* and *rock mass* and our interpretation will be guided by the conventions of soil mechanics and rock mechanics.

4.2   Soil

4.2.1   Basic soil properties

Soil consists of soil particles and voids that may be filled with water or air or both From a soil sample the unit weight γ can easily be determined by dividing the weight of the sample by its volume (e.g. the volume of the sampling cylinder). Drying a soil sample in an oven yields the dry weight of the sample from which the unit dry weight of the soil γ_d can be calculated. Oven drying yields the weight of the water that evaporates during drying from which the natural water content w can be calculated. Furthermore, the unit weight of the grains γ_g can be inferred from the dry weight of the sample and the volume of the soil particles. From these parameters, the porosity n and the void ratio e, the water content at saturation w_{max} and the saturation coefficient S_r can be deduced (Table 4.1). See Figure 4.1 for a depiction of the three-phase system. Table 4.2 lists examples of porosity and water content for different soils.

A large variety of parameters may be used to characterise soils and their agronomic value (Burt 2004, Van Reeuwijk 2006). The water retention capacity of soil is determined with a series of pressure tests. A saturated soil sample is placed in a pressure vessel on a membrane. In a first test, the pressure is increased to 1 kPa (corresponding to a nanometric pressure of 10 cm) and maintained at that level until no more water is driven out of the sample. After the sample is weighed, a second test is carried out at 4 kPa, then at 8, 16, 34.5, 100, 200 and 1500 kPa. The field capacity of the soil is defined as the suction force of the soil after several days of rainfall and refers to a pressure of 4–10 kPa. The permanent wilting point is defined as the soil water content unavailable for plants and refers to a pressure of 1500 kPa. The residual water content is determined by oven drying at 105°C after the soil sample has been pressurised at 1500 kPa. Plotting

Table 4.1 *Definition of basic properties of soils (see following pages for plasticity, consistency and activity indices)*

Quantity	Calculation
Unit weight of natural moist soil (kN m^{-3})	$\gamma = \dfrac{M}{V} = \gamma_d\,(1+w)$
Unit weight of dry soil (kN m^{-3})	$\gamma_d = \dfrac{M_d}{V} = \dfrac{\gamma}{1+w} = (1-n)\gamma_g$
Unit weight of saturated soil (kN m^{-3})	$\gamma_s = \gamma_d + n\gamma_w$
Unit weight under water (kN m^{-3})	$\gamma' = \gamma_s - \gamma_w = (1-n)\,(\gamma_g - \gamma_w)$
Unit weight of soil particles (kN m^{-3})	γ_g is determined in the laboratory
Porosity (-)	$n = 1 - \dfrac{\gamma_d}{\gamma_g} = \dfrac{e}{1+e}$
Void ratio (-)	$e = \dfrac{n}{1-n}$
Natural water content (-)	$w = \dfrac{M_w}{M_d}$
Water content at saturation (-)	$w_{max} = e\dfrac{\gamma_w}{\gamma_g}$
Coefficient of saturation (-)	$S_r = \dfrac{w}{w_{max}}$

γ_w unit weight of water (10 kN m^{-3}), M mass of soil sample (kN), M_d mass of dry soil sample (kN), M_{dT} mass of the clay fraction (<0.002 mm) of the sample (kN), M_w mass of the water of the sample (kN) (after oven drying), V volume of sample (m^3)

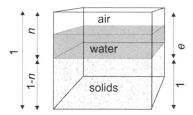

Figure 4.1 *Unit volume of the soil as three-phase system (n refers to porosity, e to void ratio)*

the water content at every pressure stage over the logarithm of the suction force yields characteristic water retention curves (Chapter 3). In addition to this method, alternative tests to determine the water retention capacity have been developed.

Organic carbon influences the mechanical behaviour of soil as well as its capacity to respond to chemical stresses. The loss-on-ignition (LOI) test is a straightforward method of measuring the amount of organic matter. The soil is first dried (usually in an oven at 105°C) until the weight remains constant, then subsequently burned in a furnace for several hours. The dry weight lost on ignition determines the organic matter content. It is calculated by dividing the difference between the sample weight before and after ignition by the initial sample weight before ignition. An alternative to the LOI test is the destruction of organic matter with chemicals. The hydrogen peroxide digestion method eliminates organic matter through oxidation. The Walkley-Black method is a more refined wet chemistry technique that involves a rapid dichromate oxidation of organic matter. As well as these physical and chemical methods, organic carbon can also be measured with qualitative tests including nuclear magnetic resonance (NMR) spectroscopy and diffuse reflectance infrared Fourier transform (DRIFT) spectroscopy. Many more methods have been proposed or are currently under development, including *in situ* tests (Schumacher 2002).

In order to measure the acidity of the soil, a dried soil sample is mixed with distilled water or a special solution (for example a salt or indicator solution). The pH-value is a measure of the acidity of the soil, also referred as water-pH. If the pH-meter

Table 4.2 *Weight, porosity and water content of typical soils (empirical values from different sources)*

	Unit weight γ (kN m^{-3})	Grain weight γ_g (kN m^{-3})	Porosity n (-)	Water content w (-)
Gravel	16–21	26–28	0.20–0.35	<0.05
Sand	13–21	26–28	0.25–0.50	0.05–0.20
Silt	16–21	26–28	0.35–0.55	0.15–0.35
Clay	16–22	26–28	0.20–0.70	0.14–0.55
Loam	18–24	26–28	0.25–0.45	0.03–0.40
Peat	10–16	15–18	0.70–0.90	0.50–8.00

shows a value lower than 7 the soil is classed as acidic, if the pH-value exceeds 7 the soil is classed as basic. A variety of kits and devices are available to determine the pH of soil samples or to measure the soil-pH directly in the field.

Soils with a pH above 6.5 may have a significant content of calcium carbonate. Alkaline soils consequently have a high buffer capacity against acid impacts. Eliminating the calcium carbonate with hydrochloric acid, while determining the volume of carbon dioxide generated i.e.

$$Ca\,CO_3 + 2H\,Cl \rightarrow Ca\,Cl_2 + H_2O + CO_2 \uparrow$$

is a common method to determine the content of $Ca\,CO_3$ in the soil. An alternative method is the Rapid Titration Method or Acid Neutralisation Method by Piper in which the sample is treated with HCL and the residual acid is titrated (IUSS 2006).

In order to detect gypsum, a soil sample is mixed with water. Acetone is added to the extract to precipitate the gypsum, which is re-dissolved in water. The Ca-concentration of this solution then serves as a measure for gypsum (IUSS 2006).

The fertility of the soil as well as its buffer capacity against contaminants is governed by the cation exchange capacity (CEC). The CEC is especially high for soil components with a large specific surface like humus and clay. Multi-layered clay minerals may have a specific surface of up to $800\,m^2\,g^{-1}$, corresponding to a CEC of up to 2000 $mmol_c\,kg^{-1}$ (millimoles of charge per kilogram). Humus may have a specific surface of up to $1000\,m^2\,g^{-1}$, corresponding to a CEC of up to 75 000 $mmol_c$ kg^{-1} (Ellerbrock 2000). A number of CEC-tests have been developed including the pH7 Ammonium Acetate test (Chapman 1965) and the $BaCl_2$-compulsive exchange procedure (Gillman 1979, Gillman and Sumpter 1986, Rhoades 1982). A straightforward way to measure CEC is percolating the soil with a solution saturated with cations. The amount of cations exchanged within the soil matrix is a measure of the CEC. The potential CEC is determined at a pH of 8, while the effective CEC refers to the natural pH of the soil. In general, the CEC decreases with decreasing pH and the capacity to buffer contaminants consequently fades. Some contaminants are, however, amphoteric, meaning that they can mobilise at either low or high pH. The salt content of the soil also influences the CEC.

The salinity of the soil is determined by measuring the electrical conductivity of the pore water. This may be achieved in the laboratory by mixing an air-dry soil sample with distilled water and measuring the electrical conductivity of the extract EC_e. By means of perforated piezometers the salinity of the soil may also be detected *in situ*. Details of the measuring procedures and further methods of measuring salinity are given in Van Reeuwijk 2006 and Burt 2004.

The sediment type is determined by a grading curve. To produce a grading curve, the soil is sieved with standard sieves beginning with the coarsest fraction progressing to finer sieves. For the finest fractions, a sedimentation analysis is carried out (Casagrande 1934). Plotting the cumulative percentage finer than a given sieve over the particle size yields the grading curve. Steep grading curves indicate well-sorted soils such as wind blown deposits (for example, loess) whereas shallow curves represent mixed soils. The grading curve ranges over boulders (> 200 mm), cobbles (60–200 mm), gravel (2–60 mm), sand (0.06–2 mm), silt (0.002–0.06 mm) and clay (< 0.002

Table 4.3 *Soil classes according to grain size*

Soil class	Acronym	Grain size (mm)
Boulders	**B**	> 200
Cobbles	**Cb**	60–200
Gravel	**G**	2–60
Sand	**S**	0.06–2
Silt	**M**	0.002–0.06
Clay	**C**	< 0.002

mm) (see also Table 4.3). Natural soils are usually a mixture of different fractions. The grading curve can be considered a fundamental tool in characterising different soil types from the point of view of soil mechanics; see Figure 4.2 for examples.

The grading curve also provides an assessment of the permeability of soil. For sandy soils, the permeability can be estimated by means of the particle size at 10 % on the percentage axis of the grading curve (Hazen 1892):

$$k = 0.0116 \, d_{10}^2$$

where d_{10} is in mm and the permeability k in m s^{-1}. Similar estimations exist for other soils. However, as will be shown later, field tests yield more realistic results in most cases.

The degree of sorting can be determined with the coefficient of uniformity U:

$$U = \frac{d_{60}}{d_{10}}$$

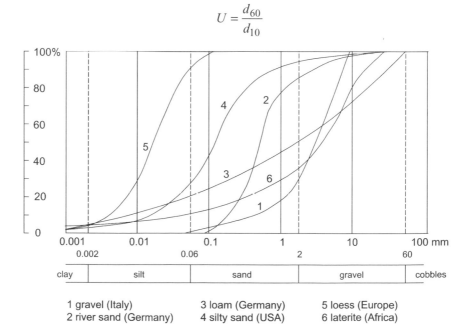

1 gravel (Italy)   3 loam (Germany)   5 loess (Europe)
2 river sand (Germany) 4 silty sand (USA)   6 laterite (Africa)

Figure 4.2 *Grading curves for typical soils*

where d_{60} and d_{10} are the particle sizes when 60 % and 10 % of the particles are passing the sieve. Uniform soils have $U < 2$ whereas well-graded soils have $U \geq 10$. Uniform soils tend to be susceptible to frost if the fraction of particles smaller than 0.02 mm exceeds 10 % while well-graded soils may be frost susceptible if the percentage of fines exceeds only 3 %. With the coefficient of curvature:

$$C = \frac{d_{30}^2}{d_{10}d_{60}}$$

soil can be further characterised, with a well-graded soil having $1 < C < 3$. The pore space of the soil is minimal for the Fuller Curve where $U = 36$ and $C = 2.25$.

The grading curve also proves useful when defining soils that can drain other soil, that is, that can be used in filters and wells. Terzaghi and Peck (1967) introduced the filter law

$$\frac{D_{15}}{d_{85}} < 4 < \frac{D_{15}}{d_{15}}; \qquad \frac{D_{50}}{d_{50}} \cong 10$$

where D_{15} and D_{50} are particle sizes of the filter material and d_{15}, d_{50} and d_{85} are particle sizes of the material to be filtered. An example of a grading curve for filter material is depicted in Figure 4.3.

In order to better define the properties of fine-grained soils, Atterberg (1911) introduced the limits of consistency. The three so-called Atterberg limits can be determined with simple index tests for which the water content of the sample is altered. The shrinkage limit w_S refers to the water content at which further drying would not lead to any further reduction of the sample volume. The plastic limit w_P corresponds to the water content at which rolling of 3-mm thick threads is not possible since the soil starts to crumble. A similar test is used in soil science to assess

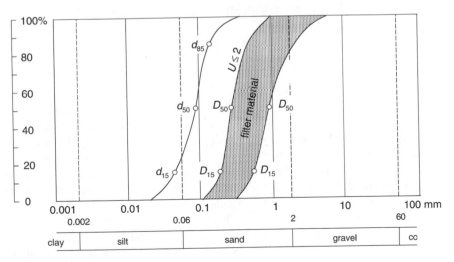

Figure 4.3 *Grading curves for filter material*

65

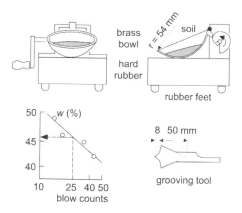

Figure 4.4 *Liquid limit device and test evaluation*

the clay content of soils. To determine the liquid limit w_L Casagrande (1932) developed a simple device consisting of a bowl that can be agitated with a special mechanism (Figure 4.4). Moist soil is fitted into the bowl and a groove is introduced with a standardised grooving tool. The bowl is agitated with sharp blows until the soil separated by the groove merges. With another sample of the same soil having a different water content the experiment is repeated. Plotting blow counts against water content yields a linear correlation; the liquid limit is given at 25 blow counts.

Although these experiments appear somewhat arbitrary, the Atterberg limits constitute important index values to characterise fine-grained soils. Consequently, and also because of the simplicity of the testing method, determination of the Atterberg limits are included in international codes of soil mechanics. Given the Atterberg limits and the natural water content w of the soil, the consistency index I_C, the plasticity index I_p and the liquid index I_L can be determined:

$$I_C = \frac{w_L - w}{w_L - w_P} = \frac{w_L - w}{I_P} = 1 - I_L$$

Casagrande (1932) plotted the plasticity index against the liquid limit to produce a plasticity chart (Figure 4.5). In this diagram, all soils above the so-called A-line are clays, whereas those below the A-line are either silts or organic clays. Organic and inorganic clays and silts can easily be distinguished by determining the content of organic matter by oven drying. Referring to Figure 4.5, CL refers to inorganic clays of low to medium plasticity, as well as gravely, sandy, silty, and lean clays. CH refers to inorganic clays of high plasticity and fat clays, ML to inorganic silts and very fine sands, rock flour, silty and clayey fine sands with slight plasticity, MH to inorganic silts, micaceous or diatomaceous fine sandy and silty soils, as well as clastic silts, OL to organic silts and organic silt-clays of low plasticity, and OH to organic clays of medium to high plasticity (Bell 1993). With the plasticity chart it is thus possible to identify fine-grained soils based on only three index tests.

66

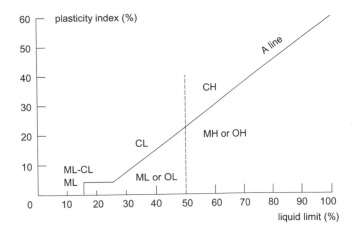

Figure 4.5 *Casagrande's plasticity chart (after Bell 1993)*

The plasticity index I_P also proves useful when assessing the swelling potential of clay. Clay minerals such as montmorillonite can store large amounts of water, a phenomenon that accompanies an increase in volume. Swelling clays may cause severe foundation problems. The swelling potential can be predicted with the activity index

$$I_A = \frac{I_P}{M_{dT}/M_d}$$

where M_{dT} is the mass of clay fraction (< 0.002 mm) and M_d is the dry mass of the sample. Clays with $I_A > 1.25$ are considered active i.e. they have a high swelling potential while clays with $I_A < 0.75$ are labelled inactive.

The compactibility of fine-grained soils is usually measured by means of the Proctor Test (Proctor 1933). In the test, soil samples of increasing water content are compacted in a mould with a certain rammer dropped from a predefined height or by means of a vibrating hammer. In order to derive the optimum water content where compaction is highest, the dry unit weight after compaction is plotted over the water content for all tests conducted. The maximum of the resulting curve (Figure 4.6) gives the highest possible density or proctor density ρ_{pr} that refers to the optimal water content w_{opt}. The dashed line indicates a compacted soil with a coefficient of saturation $S_r = 1$. Although the Proctor-Test aims at fully compacting the soil, a portion of soil pores filled with air n_a usually remains. This test is especially useful when fine soils are used as subgrade and lining material.

4.2.2 Properties of deformation and strength

If the ground is too weak, structures may suffer excessive settlement. In particular, changing ground conditions may lead to differential settlements that cause cracks and structural damage. The same effect may occur under steady ground conditions

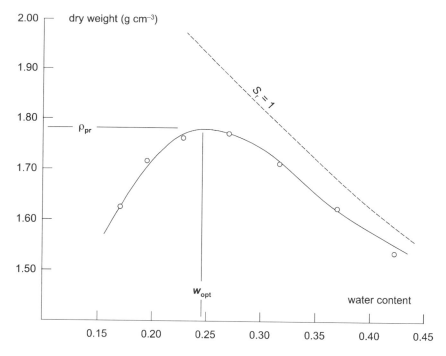

Figure 4.6 *Proctor curve with proctor density ρ_{pr} at optimal water content w_{opt}. The (dashed) line of saturation can only be obtained theoretically and refers to a soil with all pores filled with water*

if the loading varies. A simple test to predict the deformation of the ground is conducted with apparatus called an oedometer (Figure 4.7). An undisturbed sample is loaded while the deformation ε is measured to produce a stress-strain curve. From this curve a modulus of elasticity E according to Robert Hooke (1635–1703) can be inferred

$$E = \frac{\Delta\sigma}{\Delta\varepsilon}$$

where $\Delta\sigma$ is interval of stress and $\Delta\varepsilon$ is interval of strain for which the modulus of elasticity is approximated.

While the specimen is loaded deformation takes place, having both elastic and plastic components. As the sample is allowed to drain consolidation also takes place i.e. the dissipation of pore water as described by Karl Terzaghi (1925), accompanied by a deformation of the sample. In granular soil such as sand, the pore water pressure dissipates instantaneously whereas in fine soils, especially clay, the consolidation may take a long time.

During the test, the specimen is usually unloaded at least once to determine the plastic deformation at zero stress. Then, the specimen is reloaded to record its reloading behaviour. The reloading path yields a higher modulus of elasticity since

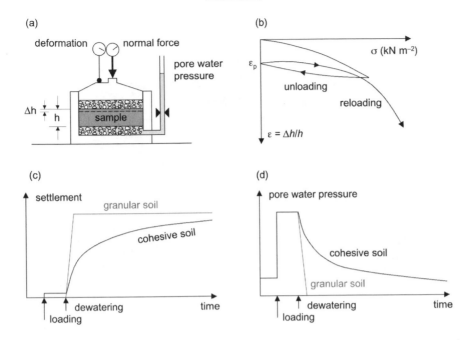

Figure 4.7 *(a) An oedometer measures deformation, used to produce (b) stress-strain curves. (c) The settlement of the soil upon loading can then be predicted, as can (d) the change of pore water pressure over time for granular and fine soils*

the ground has been consolidated already and thus behaves in a stiffer manner. As soon as the stress exceeds the original loading level, the modulus of elasticity decreases again. The stress-strain curve exposes a kink at the point where the loading exceeds the preloading. In certain regions, especially in Northern Europe and close to the Alps, glaciers that covered the land during the ice ages preloaded the soil naturally. The stress-stain curves of such soils exhibit characteristic bends from which the height of the ice cover can be calculated. The same effect can be observed in areas where erosion has taken away overlying sediments, releasing the consolidation pressure to a certain extent. Such soils are known as over-consolidated.

The strength of soil is usually expressed with friction angle and cohesion. Krey (1926) designed a simple apparatus to measure the friction angle φ. In his shear box a sample is sheared off while drainage is allowed to take place. If the shear stress τ is plotted against normal stress σ (Figure 4.8) a series of tests produces a straight line, which represents a failure criterion as formulated by the French army engineer Augustin de Coulomb (1736–1806)

$$\tau = c + (\sigma - u)\tan\varphi = c + \sigma'\tan\varphi$$

where σ is the normal stress, u the pore water pressure, φ the friction angle and c the cohesion, the latter being a property that only fine-grained soils have. (Table 4.4 lists

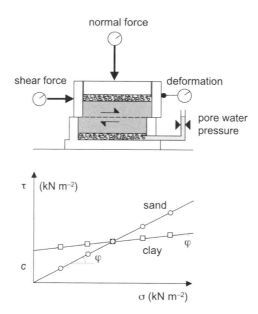

Figure 4.8 *A shear box produces τ–σ–curves to calculate friction angle φ and cohesion c of a given soil*

values of friction angle, cohesion and modulus of elasticity for typical soils.) Coulomb's failure criterion is of fundamental importance since it describes the strength characteristics of a given soil. With increasing pore water pressure, the effective normal stress σ' decreases. This aspect is certainly relevant for fine-grained soils that consolidate slowly. If fine-grained soils are loaded quickly without giving the pore water time to escape, the friction angle may not contribute to the shear strength any more. The shear strength may then only be provided by the cohesion of the soil. As well as Krey's shear box, more testing machines have been developed to measure the friction angle and the cohesion such as the ring shear apparatus and the triaxial cell.

Table 4.4 *Friction angle, cohesion and modulus of elasticity for typical soils (empirical values from different sources)*

	$\varphi\ (°)$	$c\ (kN\ m^{-2})$	$E\ (MN\ m^{-2})$
Gravel	32–42	0	100–200
Sand	32–42	0	10–100
Silt	22–35	5–20	2–15
Clay	7–32	10–60	2 15
Loam	22–40	0–25	5–20
Peat	5–30	2–20	1–5

4.2.3 Hydraulic properties

In the 19th century, the French engineer Henry Darcy (1803–58) worked in Dijon on wastewater treatment. He investigated the capacity of sand to filter wastewater, and noticed that the flow rate Q through a filter of cross-section A is proportional to the pressure difference at the points of inflow and outflow $(h_1 - h_2)$. He also noted that the filter length l influences the flow rate (Figure 4.9). He concluded that the speed of flow v can be related to the length of the filter l, the hydrostatic gradient i and a material property k that depends on the type of filter material employed, commonly referred to as the coefficient of permeability, or simply permeability, in civil engineering:

$$\frac{Q}{A} = v = k\frac{h_1 - h_2}{l} = k\,i$$

Darcy's law (1856) is fundamental for the hydraulic characterisation of soils. The permeability k is a soil-specific parameter, for which typical values are listed in Table 4.5. It is small in fine-grained soil but large in coarse material. The permeability can be measured in the laboratory with the apparatus developed by Darcy or in the field by means of infiltration and pumping tests.

Since the ground consists of layers of sediments, the permeability varies with depth. Layers of sediments have a higher permeability in the horizontal direction since in the vertical direction the lowest permeable layer dominates the flow.

Furthermore, it has to be kept in mind that many sediments expose fractures (e.g. Figure 4.10) due to geological processes such as preloading or tectonic movements.

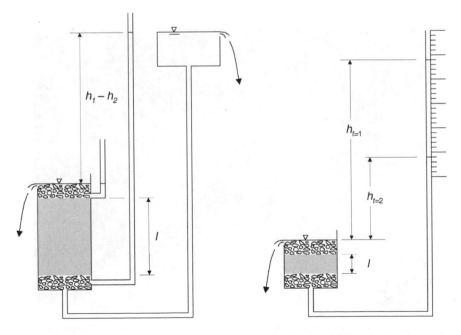

Figure 4.9 *Darcy's experiment with constant hydraulic head and falling hydraulic head*

71

Table 4.5 *Permeability of typical soils (empirical values from different sources)*

	$k\ (m\ s^{-1})$
Gravel	$2 \times 10^{-1} - 1 \times 10^{-2}$
Sand	$1 \times 10^{-3} - 1 \times 10^{-5}$
Silt	$1 \times 10^{-5} - 1 \times 10^{-9}$
Clay	$1 \times 10^{-7} - 1 \times 10^{-11}$
Loam	$1 \times 10^{-5} - 1 \times 10^{-9}$
Peat	$1 \times 10^{-5} - 1 \times 10^{-8}$

The hydraulic properties may be governed by these fractures, an aspect that is often not taken into consideration when the properties of a geological barrier are considered, for example, when searching for an appropriate site to build a landfill.

In general, field tests are more reliable since they affect a much larger part of the ground. They are, on the other hand, much more expensive and are usually only carried out during a site investigation campaign when borings have to be sunk to investigate the ground. These borings may subsequently be utilised as wells to determine the permeability *in situ*. Associating the law of continuity with Darcy's law yields, in connection with geometric considerations, the Special Well Formula as derived by Dupuit in 1863 and modified by Thiem in 1906. For a well confined at the bottom by a perfect aquiclude (a layer with zero permeability) and steady flow

Figure 4.10 *An area of fractured clay (about 10 m^2) at the Münchehagen Hazardous Waste Landfill, Germany (photograph courtesy of Pickel)*

conditions (constant water extraction from the well), the amount of water extracted per unit time Q (m^3 s^{-1}) is related to the permeability k (m s^{-1}) by

$$Q = \frac{\pi k (H^2 - h^2)}{\ln R - \ln r}$$

where H, h, R and r define the geometry of the depression cone (Figure 4.11). Since the distance R at which the depression cone merges with the original groundwater table is not know when conducting a single well test, it has to be estimated (Sichardt 1928):

$$R = 3000 s \sqrt{k}$$

or (Strzodka 1977)

$$R = 575 s \sqrt{H k}$$

where s is the depth of the depression cone (Figure 4.11). Since k is required to estimate R, the permeability has to be calculated iteratively. If piezometers are installed around the well to observe the depression cone, the Special Well Formula simplifies to the General Well Formula:

$$Q = \frac{\pi k (z_2^2 - z_1^2)}{\ln x_2 - \ln x_1}$$

where x_1 and x_2 are the distances to the piezometer readings z_1 and z_2 (Figure 4.12). In this case, no estimation of R is necessary and the permeability k can be determined directly.

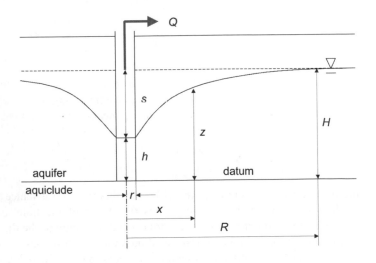

Figure 4.11 *A single well test with a depression cone around a well*

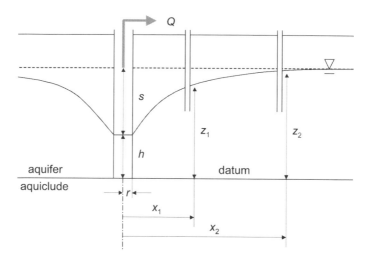

Figure 4.12 *A well test with observation piezometers*

Another way to assess the permeability of the ground with a single well is to change the water level within the well by filling up or pumping and subsequently observing the time taken until the original groundwater table is restored. Fill up and drawdown tests can be run without the necessary equipment to ensure a constant pumping rate. For this well test, the change of the well level Δs is recorded for each time interval Δt to derive a mean water level change s_m. With the effective inner radius r_f of the well (the reduced radius due to an obstacle such as a pump) the permeability is calculated as:

$$k \cong \pi r_f^2 \, \Psi \frac{\Delta s}{\Delta t} \frac{1}{s_m}$$

The parameter Ψ describes the specific geometry of the test.

The permeability can be assessed without sinking a well, using simple infiltration tests. In order to conduct such a test, a shallow circular hole is dug and filled with water to observe the water level drop over time. The permeability is estimated using

$$k \cong \frac{d}{28} \frac{\Delta s}{\Delta t} \frac{1}{s_m}$$

where d is the diameter of the hole (Figure 4.13). For this test, the distance to the groundwater table should be at least seven times the depth of the hole and the maximum grain size of the soil particles should not exceed one-tenth of the diameter. However, since the water flows into an unsaturated medium, this test does not give a true measure of permeability and the outcome can only be considered a rough estimate.

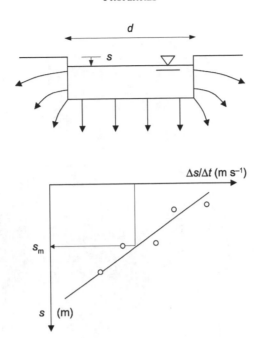

Figure 4.13 *An infiltration test and evaluation*

4.3 Rock

4.3.1 Rock and fractures

In comparison to soil, rock is massive rather than granular and only separated by sets of fractures (see Figure 4.14, for example) that render rock discontinuous. Rock is consequently made up of intact rock blocks that make up the rock mass. The physical properties of intact rock are governed by the minerals, distribution and cementation, whereas for rock mass the distribution and characteristics of the discontinuities are decisive.

In order to map and analyse these discontinuities a joint survey is carried out, aimed at identifying all the different sets of discontinuities. In sedimentary rock, the bedding planes that reflect the fluctuations of the sedimentation are also interpreted as discontinuities. By means of a geological compass the dip ϑ (the inclination) and the azimuth of the dip direction α are measured for a representative number of samples of each set of discontinuities identified. The measurements are interpreted as vectors (α, ϑ) and plotted on a diagram referred to as a Schmidt Net (Schmidt 1932, Sander 1948, 1950), an azimuthal Lambert projection, where they are represented by so-called pole points p' i.e. the points of intersection of the vectors with the unit half-sphere projected as a Schmidt Net (Figure 4.15). They are interpreted statistically or by means of an eigenvector analysis to form clusters and bands that characterise the fracturing of the rock mass. A block diagram depicting different types of discontinuities and their interpretation with the Schmidt-Net can be seen in Figure 4.16.

75

Figure 4.14 *Buildings on fractured rock in Ronda, Spain*

For each set of discontinuities, further aspects relevant to the assessment of their geotechnical behaviour are recorded. They include mean estimates of the following measurements:

- Spatial extension: the length of the discontinuity as measured along the outcrop (m)
- Persistence: the sum of outcropping fracture traces belonging to a single fracture divided by the measuring line along which they are recorded
- Aperture: the opening of the discontinuity (cm)

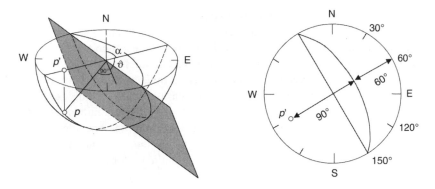

Figure 4.15 *The discontinuity measured in the field (grey plane) intersects the lower half-sphere. The trace is projected to the Schmidt Net to form a great circle. The vector perpendicular to the discontinuity intersects the lower half-sphere to form a pole point p that can also be projected to the Schmidt Net. A great circle can thus be reduced to a single point p'*

- Degree of weathering: the alteration of the intact rock due to weathering processes
- Habitus: the intensity of surface irregularities along the fracture based on tactile and visual inspection (smooth, rough, undulating, zig-zag)
- Intensity of fracturing: the reciprocal of the number of discontinuities cutting a measuring line (m^{-1}) (Stini 1922).

A similar measure to quantify the fracture intensity from a rock core extracted from a boring is the Rock Quality Designation (Deere 1964). It is defined as the

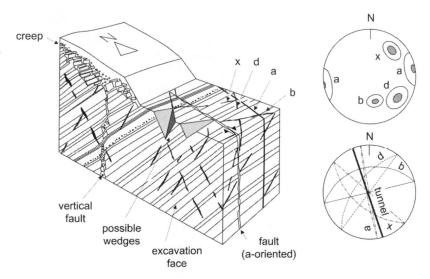

Figure 4.16 *A block diagram of rock mass as mapped for an excavation site. Indicated are the different sets of discontinuities (a, b, d, x) as well as possible wedges that could slide into the excavation. The discontinuity pattern is indicated with Schmidt nets: upper, statistical interpretation for the identified discontinuity sets and lower, great circles as defined by the mean pole points*

percentage of rock cores longer than 10 cm per core run. For example, given a drilled length of 1 m and the sum of core fragments exceeding 10 cm as 0.85 m, the RQD is 85 %. Rock fractures introduced by the drilling process are not counted.

The intensity of fracturing usually decreases with depth. However, close to a fault system the intensity of fracturing also increases. In fact, faults may govern the mechanical and hydraulic behaviour of a given rock mass. Only experienced geologists are able to conduct a survey of the pattern of discontinuities consisting of faults (see Figure 4.16), joints, bedding planes, schistocities and cleavages. The evaluation of the discontinuity pattern gives an insight into the inner architecture of the rock mass that controls its mechanical and hydraulic performance.

4.3.2 Properties of deformation and strength

The mechanical behaviour of intact rock and rock mass differs considerably. The elastic properties of an intact rock sample, for example a core from a boring, can easily be determined by means of a uniaxial compressive test. With increasing axial stress the axial strain increases in an almost linear way, producing a modulus of elasticity E (see Section 4.2.2).

When loaded until failure, the uniaxial compressive strength σ_u of the rock sample is measured (Figure 4.17). The strength of intact rock is important, for example, when judging the quality of building stones. A simple field test to assess the uniaxial compressive test is the point loading test, for which a sample is fitted between two cones that are loaded with a hydraulic hand pump (Figure 4.18). The uniaxial compressive strength σ_u is calculated as

$$\sigma_u = K I_S$$

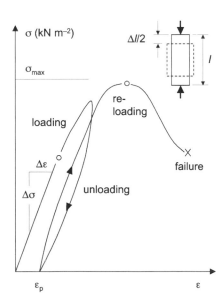

Figure 4.17 *Stress-strain diagram as derived from a uniaxial compression test*

Figure 4.18 *Point load test device*

where K is an empirical factor and I_S, the point load index, is given by

$$I_S = \frac{F}{d^2}$$

where d is the distance between the cones and F the point load at failure.

The linear correlation between uniaxial strength σ_u and point load index I_S is demonstrated in Figure 4.19.

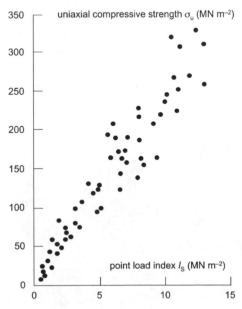

Figure 4.19 *Correlation between the point load index I_S and the uniaxial compressive strength (for 54-mm cores, after Hoek and Bray 1981)*

79

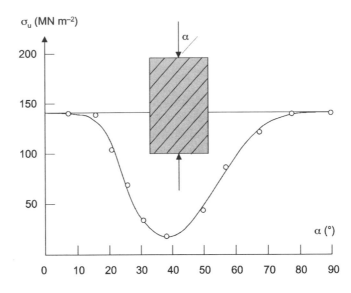

Figure 4.20 *Uniaxial compressive strength of rock mass as a function of azimuth dip direction (simplified)*

The uniaxial compressive strength can also be correlated with the rebound hardness as measured with the Schmidt hammer (Figure 4.21), which is routinely used to test the strength and quality of hardened concrete. An example correlation is presented in Figure 4.22 for German Triassic sandstone.

The deformation and strength characteristics of rock mass depend on the intensity of fracturing and the orientation of the fractures. With increasing fracturing the strength decreases and the deformation becomes more plastic. The strength of fractured rock is highest when the discontinuities are oriented perpendicular to the axis of loading (Figure 4.20). With increasing inclination, the discontinuities begin to shear off. The shearing resistance along a discontinuity depends on the mobilised

Figure 4.21 *Schmidt hammer*

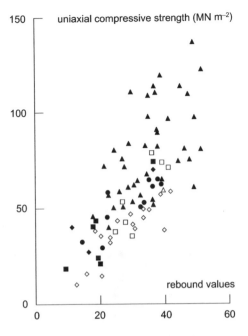

Figure 4.22 *Correlation between uniaxial compressive strength and Schmidt rebound hardness for German Triassic sandstone, where the four ornaments represent different locations (after Prinz 1997)*

friction angle φ_m, which is a function of the joint roughness coefficient (JRC), the uniaxial compressive strength of the surface of the discontinuities σ_j (as measured, for instance, with the Schmidt hammer) and the basic friction angle φ_b i.e. the friction angle along the smooth (sawed) rock surface (Barton 1973, 1986)

$$\tau = \sigma \tan \varphi_m = \sigma \tan \left(\varphi_b + \text{JRC} \log \frac{\sigma_j}{\sigma} \right)$$

(see Figure 4.23 for examples of τ–σ curves).

With the JRC taken from model profiles, this empirical law only allows the shear behaviour of a fracture to be approximated.

4.3.3 Hydraulic properties

The permeability of rock mass is governed by the pattern of discontinuities. It increases with the density of open fractures and increasing apertures (see Figure 4.24). However, there appears to be a hierarchy of water bearing fractures, starting from small fissures that lead water to larger fractures and eventually to dominant flow paths. Consequently, dominant open fractures govern the water movement. Moreover, the intersection of fractures may play an important role since they channel water much more effectively than single fractures, as was proven at an outcrop in Spain (Figure 4.25) (Bruines and Genske 2001, Bruines 2003, Genske 2003).

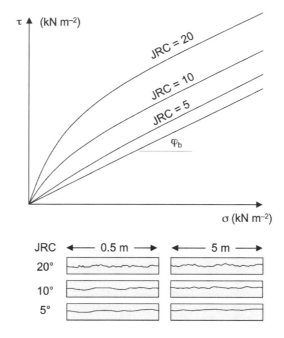

Figure 4.23 *τ–σ–curves for rock fractures*

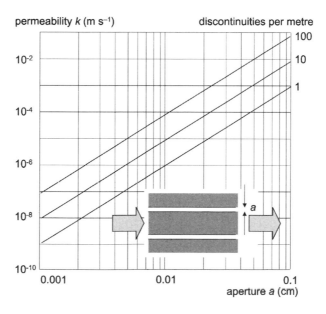

Figure 4.24 *Estimation of bedrock permeability in the direction of observed discontinuities based on fracture intensity and aperture (after Hoek and Bray 1981)*

(a)

(b)

Figure 4.25 *(a) Exposed bedding planes of a limestone variety (a biosparite of the Middle Lias/Carixian) near Granada, Spain; (b) due to the arid climate, fossil flow channels that developed along the intersection of discontinuities have been preserved*

Close to the surface, the discontinuities are often locked with loam from weathering processes, causing a decline of the overall permeability. It is therefore difficult to judge the permeability of the rock mass from the surface especially since the fracture density usually decreases with increasing depth as mentioned previously.

An assessment of the *in situ* permeability of fractured rock can be made by sealing off an uncased section of a borehole with inflatable packers and pumping water into the sealed off section. For a packer test (Figure 4.26), the water pressure is increased while the water intake is measured and subsequently slowly released. The water intake is measured in $l\,min^{-1}$ (litres per minute) and metre of borehole or *Lugeon*, after

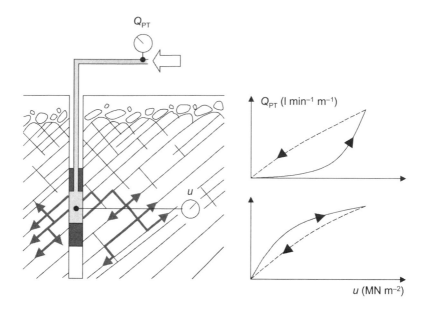

Figure 4.26 *A packer test and test results. The water intake Q_{PT} plotted over the injection pressure u characterises the hydraulic behaviour of the rock mass*

the French geologist Maurice Lugeon (1870–1953) (Lugeon 1933). 1 Lugeon is equivalent to 1 l min^{-1} and metre of borehole at 1 mPa. A slow water intake that later accelerates indicates either erosion of plugged fractures or an increase of conductivity due to the opening of cracks. A rapid water intake that later slows implies either the transition from laminar to turbulent flow or plugging of fractures. Heitfeld (1979) derived the correlation between the water intake Q_{PT} and the permeability of fractured rock based on field observations at dam projects in the Rhenish Massif (Germany):

$$k \cong (3.3 Q_{PT}^2 - 1.3 Q_{PT}) \times 10^{-8}; \quad 1 < Q_{PT} \le 25$$

where Q_{PT} is measured in l min^{-1} (or m of borehole at 0.5 mPa) and k in m s^{-1}. Nevertheless, it must be stated that the assessment of the permeability of rock mass is a complex problem and only estimates are possible.

II Degraded land

In Part II, we discuss processes of ground denudation due to erosion. Thereafter, aspects of chemical degradation are introduced, notably contamination, salinization, acidification and solidification. Finally, features of physical degradation are developed, including surface sealing, compaction, subsidence and water logging.

5 Erosion

Types of erosion, namely water and wind, are presented and the scope of the problem is defined. The causes of erosion are investigated and approaches to modelling and predicting erosion phenomena are introduced. Finally, the consequences of erosion including on- and off-site effects are discussed.

5.1 The *Bishnoi* sacrifice

In 1730, the Maharaja Abhay Singh of Jodhpur, India, needed firewood to burn lime to construct a fort close to the village of Khejadli. When the woodcutters arrived they were stunned to see the people of the village—men, women and children—embracing the trees in order to protect them from being cut. The woodcutters tried to convince the villagers to obey the imperial demand, but whatever argument was put forward they wouldn't give in. The Maharaja ordered the army to restore order—in vain. Eventually, he commanded his army to behead the obstinate villagers so that the trees could be cut. On that day, 363 villagers died in a desperate attempt to save their forest.

The villagers belonged to the *Bishnoi* community, a small ethnic group believing in twenty-nine (*Bees + nau*) tenets as formulated by Guru Jambheshwar in the 15th century, a well-known social reformer in those days. One of the commitments prohibits killing wild animals, and another forbids cutting green trees and vegetation. Today, *Bishnoi* communities still subsist in the ecologically fragile region of northwest Rajasthan, Haryana and Punjab. It is due to their effort that the vegetation cover is still intact and wildlife is present around their villages. In their ecological awareness they can be compared with the *Kuna* of Panama, the *Tuarege* of Niger and the *Kayapo* of Brazil (Qureshi 2004).

These efforts are, unfortunately, rare exceptions. Many ancient cultures collapsed because of deforestation, prompting soil depletion and consequently famine, resulting in a destabilisation of the affected societies. Eventually most of these cultures disappeared. Examples are the vanished cultures of the Easter, Pitcairn and Henderson Islands in the Pacific, the Maya civilisation of the Yucatan Peninsula of Mexico and Guatemala, the Anasazi Culture of North America and the abandoned settlements of the Vikings on Greenland (Diamond 2006).

Once plants are removed, soil loses its protective cover and the nutrient rich A-horizon is successively denuded. Rain washes away the active soil horizons and wind blows them out, a process called erosion with the distinction made between water or *fluvial* erosion and wind or *aeolian* erosion. Eroded sediments cover land and watercourses, damaging terrestrial and aqueous ecosystems as well as man-made structures (for example, Figure

5.1). After erosion has consumed the active soil horizons, it becomes difficult for plants to re-establish and flourish. The soil becomes sterile and loses its ability to support life, a process referred to as *desertification*. Desertified land is difficult to restore. Erosion has therefore been a subject of extensive research since the early 19th century.

The recent deterioration of the Earth's climate (as described by the Intergovernmental Panel on Climate Change, for example) further accelerates erosion processes. For example, the escalation of storm events of increasing intensities erodes soils already denuded by clear-cutting and farming, at a pace that worries politicians as well as experts. When the tropical cyclone Gafilo hit the island of Madagascar in March 2004, astronauts aboard the International Space Station observed a massive sediment plume flowing into the Betsiboka River estuary and the ocean. The sediments were eroded from the former rainforest that had been logged and cleared many years previous (UNEP 2005).

5.2 Scope of the problem

Human-induced soil degradation may be described as breaking the balance between the attacking forces of the climate and the natural resistance of the terrain against these forces. Human intervention introduces a fall in the present and possibly future capacity of soil to support life (Oldeman *et al.* 1991). If our natural environment was

Figure 5.1 *The harbour of the ancient city of Ephesos (Turkey) has been silted-in due to deforestation many centuries ago. The 'Harbour Road' now leads to a marshy terrain far away from the sea*

in balance, background denudation would remove soil at about the same rate as it was formed. Our environment is, however, not in balance. George Perkins Marsh warned his compatriots in the young American Republic not to repeat the mistakes made in the old world. In his book *Man and Nature* (1864) he predicted the devastating effects of erosion due to human impact. More books followed (Bennett 1938, Jacks and Whyte 1939, Lal 1994, Morgan 2005)—alas without success. The routine of land use remained unchanged, and in many cases worsened.

On farmed land, erosion is highest during the period between ploughing and crop growth beyond the seeding stage (Morgan 2005). At that time, the denuded soil is not or only barely covered and thus exposed to wind and rain. While erosion takes place, the soil profile along the cultivated slopes diminishes. Water carries the soil to the valley floor where it accumulates as *colluvium* (Figure 5.2). Wind carries fine particles to distant areas where they typically form *loess* deposits. The first traces of human-induced erosion can be found in Neolithic times i.e. from 4000 BC onwards (Lüttig 1960, Wildhagen and Meyer 1972, Mücher 1986, Bork 1989). With increasing populations, the demand for farm and grazing land increased, as did the demand for firewood for warmth, cooking and lumber to construct buildings and structures. Enormous amounts of charcoal were burned in the once extensive forests. The soils deprived of forest became vulnerable to the forces of erosion that swept over Europe in "destructive waves" (Richter 1976). The first reached Europe between 1313 and 1350, after annual erosion rates had already increased to about 10 ton ha^{-1} (Bork 1989, Bork *et al.* 1998) (Figure 5.3). Extreme weather events such as the

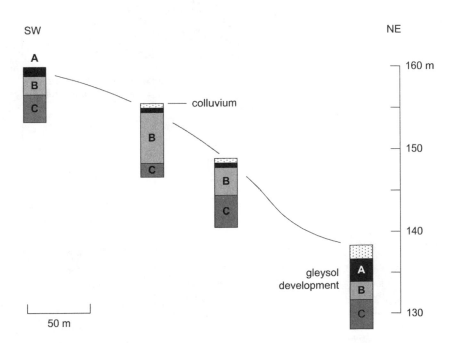

Figure 5.2 *Catena of an eroding slope. At the upper slope the profile diminishes while new soil is formed at the toe of the slope (simplified, after Lessmann-Schoch et al. 1991)*

rainstorm of the 21st of July 1342 caused annual erosion rates to peak at 2250 ton ha⁻¹. While this period of rain and erosion was coming to an end, the Black Death devastated the land, killing one-third of Europe's population. Vast areas became abandoned and forest re-established, prompting a fall in erosion. With the rehabilitation of population numbers and the introduction of the three-field cropping practice, with one field in fallow at any one time, erosion rates increased to about 25 ton ha⁻¹ yr⁻¹. However, from 1749 to 1800 the weather deteriorated again and annual erosion rates reached 160 ton ha⁻¹. The fact that a higher proportion of land was under grass and trees, and better land management practices had been introduced like terracing and contour ploughing, prevented even higher soil losses. At around 1800, annual soil loss stabilised at about 20 ton ha⁻¹, but has increased again in recent years due to land consolidation with larger fields, removal of terraces and grass strips and levelling of land (Morgan 2005).

Since the time when humans began to cultivate crops, human-induced soil erosion has severely degraded about 200–300 Mha i.e. up to 10 times the size of the UK or almost one-fifth of the planet's arable land (Myers 1988, Lal 1988, FAO-AGL 2000, Bot *et al.* 2000). This land is characterised by a greatly reduced biodiversity and the

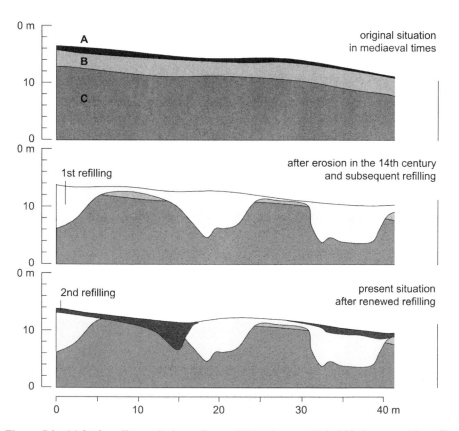

Figure 5.3 *Multiple gullying of a loess slope at Rüdershausen, Eichsfeld, Germany. The gullies formed during episodes of heavy rainfall and were subsequently refilled (after Bork 1983)*

90

incapacity to produce food. The Food and Agricultural Organisation FAO estimates the global loss of productive land through erosion (water and wind) is $10 \times 10^6 \, \text{ha yr}^{-1}$.

In Europe, three major erosion zones have been identified (EEA 2000). In the southern zone, seasonal strong rains reduce the typically thin Mediterranean soil profiles. The harsh climate, the rough topography, unsustainable farming practices and frequent forest fires further increase the soil loss. In the northern zone that stretches along France up to Russia, the soil is characterised by a high silt content, which makes it more vulnerable to water and wind erosion. This zone is also referred to as the loess belt. The eastern zone covers Central and Eastern Europe where the land has traditionally been managed by large *kolkhozes* in a way that in most cases accelerates both water and wind erosion. For example, in the Czech Republic and Bulgaria more than 50 % and 70 % of the farmland respectively is affected by water erosion (Montanarella 2003). The European Environmental Agency EEA estimates that about 16 % of European soils are affected by water and wind erosion (EEA 2003). The annual erosion rates may well exceed 25 ton ha^{-1} yr^{-1} in many regions (EEA 2000). However, during individual storms 20–40 ton ha^{-1} of topsoil may be eroded, in extreme events even up to 100 ton ha^{-1} (Jones *et al.* 2003). These storms occur every two or three years. It is estimated that a soil loss of 1 ton ha^{-1} corresponds to a reduction in soil thickness of about 0.1 mm (Alexander 1988, Morgan 2005).

Even in Switzerland, a country known for its environmental awareness, up to 2 ton ha^{-1} are eroded every year in the Central Lowlands, although they are permanently vegetated. Heavy rain may, however, increase the loss of soil to up to 40 ton ha^{-1} (BFS 2002). In the United States, soil is eroded at a rate seventeen times greater than it is formed. 90 % of US cropland is currently loosing soil above the sustainable rate (IUGS 2005). The damage to the US economy is estimated to be in the order of 30–45 billion US$ per year (Morgan 2005). In parts of Asia, Africa and South America, soil erosion rates are estimated to be twice as high as this. An estimation made for 8 countries of the South Asian Region revealed that about 7 % of the economic value of agricultural production is consumed by land degradation, without taking into account offsite effects like siltation of reservoirs (Young 1998). In China, nearly 30 % of the land area is affected by desertification due to overfarming, overgrazing and deforestation, causing annual economic losses of some 6 billion US$ (UNEP 2004).

Soil is not only denuded by erosion however. The upper active horizons of the soil profile are also removed voluntarily in the course of human development. For example, soil is removed to reach resources like sand and clay or to reach the bedrock to dig for building stones, minerals and coal. Moreover, soil is removed to create appropriate ground for buildings, roads and other constructions. The A-horizon with its loose organic surface, the B-horizon with its dynamics of pedogenesis, the interaction between the two and their inhomogeneity make them unsuitable to support structural loading. A sterile, homogeneous and well-compacted dense soil with a high modulus of elasticity and a high shear resistance is desired (Chapter 4). Only on these soil structures can building foundations be safely constructed.

The expansion of human settlements is inevitably accompanied with the denudation of land. However, as soon as constructions are erected and roads are paved, the denuded land is covered and loss of soil is halted, whereas on farmed land the erosion continues and accumulates to disturbing figures. On the other hand, as soon as constructions are not used anymore and become abandoned, the terrain falls derelict and brownfields develop. Derelict terrain is prone to erosion again, especially if the constructions are dismantled and no effort is made to stabilise and re-vegetate the ground. Water and wind erosion may mobilise contaminants that are washed into rivers and lakes or blown into the surrounding areas.

Sedimentation of the Guadalquivir-Delta, Spain

Christopher Columbus left Palos de la Frontera on the 3rd of August 1492 with three caravels, sponsored by Ferdinand II and Isabella of Castilia, to find the sea route to India. His journey was to be in vain, as we know. The quay from where he departed still stands, as does the well he utilised to fill the freshwater tanks of his ships. Today, however, the harbour is filled with sediments that were eroded from the Sierra Morena (Figure 5.4) into the Golf of Cadiz. From Phoenician times, the hinterland has been cultivated and exploited by farming and mining. Vast areas were clear-cut to satisfy the growing demand for wood, prompting a dramatic increase in erosion. Mining is still ongoing today in the Rio Tinto Mining District. Due to human impact, arable land has been washed into the sea, creating a quickly expanding delta while the inland has become more and more barren. Efforts to reforest the hills are currently underway, however (Figure 5.5).

Figure 5.4 *The port from where Columbus set off is filled with sediments of the Guadalquivir, which now reaches the Gulf of Cadiz way downstream: on the left, the terrain surrounded by bushes is the former harbour basin, and the meadow on the right marks the position of the former quay*

Figure 5.5 *Reforestation has commenced to reduce water and wind erosion*

5.3 Causes

Erosion is caused by a number of factors, most of which are man-made (Barrow 1994, Goudie 2000, Morgan 2005). The problem begins with the removal of the vegetation cover which protects the soil from the impacts of rain and wind and the activities of man. Human-induced erosion may be due to

- clearing
- farming
- lifestock herding

i.e. factors directly associated with turning natural land into pasture and making it arable, as well as

- urbanisation
- tourism
- poverty

i.e. factors directly associated with transforming land to accommodate humans.

Clear-cutting appears to be one of the major factors triggering erosion. Introducing monocultures on cleared land, for instance maize or cotton, considerably increases the problem. In recent years, the introduction of genetically modified (GM) crops has aggravated the problem. The fast crop yields promised by the biotechnical industry seduces farmers to deforest entire regions. Moreover, soils not suitable for cultivation are farmed and large amounts of fertilisers, pesticides and insecticides are applied (Chapter 6). The promotion of transgenic crops by only a few multinational companies leads to extensive erosion problems in regions that were previously stable and ecologically sound. Moreover, roads, railways and waterways are constructed to make these areas accessible, further degrading the ground.

Grazing of cattle and other livestock reduces the vegetation cover and makes soil susceptible to erosion. In addition, animal trampling both compacts and dislodges soil. The effects of grazing and ground distortion enhance one other, leading to considerable soil loss.

Minor forest fires, being part of the natural dynamics, have only a limited effect on soil degradation. However, from the beginning of agriculture, man has deliberately burned forests to produce arable land. With increasing populations, the practice of slash-and-burn has progressively reduced soil profiles all over the world. Forests are still burned today, in many cases illegally, not only for farmland but also for urban development. Burning the vegetation cover also means the destruction of the protective litter cover and the damaging of the root system. The resistance to erosion is consequently exhausted. In Europe, the Mediterranean region is most affected by forest fires. Every year, more than 50 000 fires are recorded—almost all caused by humans. An average of 5000–8000 km^2 are burnt, an area comparable to the islands of Crete or Corsica and equivalent to 1.3–1.7 % of the Mediterranean forests (Rokos and Kolokoussis 2004, UNEP 2003). Since the current climate change provokes longer and more frequent droughts and stronger winds, the situation is expected to deteriorate. Moreover, the intensity and frequency of rainfall has significantly increased during the rainy season, provoking a rapid escalation of soil erosion.

Stripping off the upper soil horizons to create suitable building ground denudes large areas. High erosion rates are observed during construction, when soil is excavated and machinery moved. Wolman and Schick (1967) recorded that during a single year of construction, the rate of erosion is equivalent to many decades of natural erosion and many years of agricultural erosion. As soon as the construction work comes to an end, the roads laid and the gardens cultivated, the erosion rate returns to normal (Goudie 2000).

With the introduction of holiday resorts all over the world, soil erosion is reported from regions previously unaffected. Since tourists are inclined to spend their vacations in a natural environment differing from the urban environment they are used to, they destroy nature while actually seeking it. In tourist regions, the natural appearance of the environment is usually abbreviated to permit a maximisation of tourist numbers. The resorts bordering the Mediterranean Sea, for instance, absorb about 30 % of the international tourist streams (EEA 2000). But even those who seek a natural environment away from the commercial holiday resorts damage the soil by intruding the countryside with all-terrain vehicles and mountain bikes. Skidoos and skiers degrade the fragile alpine and tundra soils and make them prone to erosion. The recent introduction of quad bikes significantly increases the damage already done. Even hikers may damage the soil by following walked-out tracks that may collect rainwater and eventually develop into erosion gullies.

5.4 Types of erosion

5.4.1 Fluvial erosion

Fluvial erosion is the detachment of soil due to the impact of water. The following types of fluvial erosion are distinguished:

- Splash erosion i.e. loosening and detachment of soil particles due to the impact of raindrops on bare soil (Ellison 1944). The soil remains on site however, and moves slowly with the inclination of the slope. Splash erosion underneath trees is the perfect example for demonstrating the complexity of erosion. The crown reduces the intensity of the rainfall, but rainwater accumulates on the leaves to form bigger droplets, increasing the impact. However, the tree provides shade as well as nutrients from falling leaves, thus stimulating the development of a layer of litter (L-horizon) and a vegetative cover that in turn buffers the impact.
- When the rainfall intensity exceeds the infiltration rate or when the infiltration is hampered because the soil is already saturated, water may leave the area as surface runoff. The two situations which can develop are referred to as infiltration excess runoff and saturation excess runoff. If the runoff extends across the surface as unconfined flow, sheet erosion takes place.
- While sheet erosion takes place, small channels or rills may develop, giving rise to rill erosion.
- Larger channels or gullies may collect great quantities of runoff, thus giving rise to gully erosion (for example, Figure 5.6).

As erosion processes evolve, a hierarchy of rills, microrills and gullies develops that removes soil from its original place and transports it to rivers, lakes and eventually into the sea.

Figure 5.6 *An eroding slope at Yosemite National Park, USA; erosion has destroyed all vegetation and has provoked considerable gullying*

The amount of soil lost by erosion depends on many factors including land use, soil cover, inclination and other parameters. Assessing soil loss is therefore considered to be a difficult task. Until recently, the dimension of the problem could only be judged indirectly by analysing the rate of sedimentation in lakes and along the continental shelves. These investigations indicate, for example, that sedimentation increased significantly when Mesolithic and Neolithic people started to clear the forest to make land arable. Sedimentation increased further when agriculture intensified and the iron plough was introduced during the Iron Age. With increasing population, the sedimentation rates have continued to accelerate with peak erosion events as previously mentioned.

With the beginning of industrial farming in the Corn Belt of the United States, fluvial erosion developed into a major problem. After World War II, the situation became even worse. In 1946, erosion specialists held a workshop in Ohio to define the major factors governing fluvial erosion and to derive an equation to quantify erosion intensity. After nine years of data collection, the Universal Soil Loss Equation (USLE) was published together with a handbook to explain its application (Wischmeier and Smith 1965, 1978, Meyer 1984). The concept of the equation was accepted as a common working tool in many other countries and the Revised Universal Soil Loss Equation (RUSLE) was developed (Renard *et al.* 1991, 1994, 1997). The mean annual soil loss A_f is calculated as

$$A_f = I_f \, E_f \, S_f \, C_f \, P_f$$

where I_f is erosivity, the impact provoking fluvial erosion. It reflects rainfall and runoff events and depends on the frequency and intensity of storms; values are listed in manuals for different regions. E_f is the erodibility, or the susceptibility of the soil to fluvial erosion. Clay-rich soils have a low E_f value (0.05–0.15) since they resist detachment because of their cohesion. Loamy soils have moderate E_f values (0.25–0.40), whereas soils with high silt content are easily detached and thus have high E_f values (>0.40). Coarse sandy soils, on the other hand, have low E_f values due to typically high infiltration rates reducing the runoff. Organic matter increases the cohesion and thus lowers the E_f value. S_f is a spatial component, listed in tables, dependent upon the slope length and steepness. C_f is a measure of the soil cover deviation from standard conditions (a clean-tilled field). It is a function of vegetation, roughness of soil profile, farming and soil management techniques. The erosion-control practice factor P_f reflects the type and frequency of annual mechanical disturbance routines (contour ploughing, strip cropping etc.). In most publications, A_f, I_f, E_f, S_f, C_f and P_f are given as E, R, K, L (or S), C and P respectively, although the apparent similarity to the *Wind Erosion Equation* (WEQ) below would suggest matching parameters.

The RUSLE equation represents an empirical approach to forecasting and modelling soil loss due to fluvial erosion. In recent years, the RUSLE has been extended to the assessment of soil loss on construction and derelict sites, mine spoil and reclaimed land (Toy and Foster 1998). The calculation can be carried out online via a number of web portals (for example IWR 2005). The Soil Loss Estimator for Southern Africa (SLEMSA) (Elwell 1978) is another empirical model adapted to the countries of Southern Africa whereas the Morgan, Morgan and Finney Method (Morgan *et al.* 1984, Morgan 2001) has been applied to Indonesia, Nepal and Mediterranean Europe.

In order to assess the spatial distribution of runoff and soil losses during individual storm events, as well as total soil loss and total runoff, process-based models have been introduced. Examples are the Water Erosion Prediction Project (WEPP) (Lane and Nearing 1989), the Griffith Universal Erosion Sedimentation System GUESS (Rose *et al.* 1983), the EROSION-Approach (Schmidt 1991) and the European Soil Erosion Model EUROSEM (Morgan *et al.* 1998). Process-based models are predicated on physical boundary conditions and take into consideration the conservation of mass and energy. They have proven to be of particular interest when quantifying off-site effects. An overview including advantages and disadvantages of these models is given in Schmidt (1998) and Morgan (2005).

In recent years, attempts have been made to combine soil erosion prediction models with remote sensing techniques based on geo-information systems (GIS). Different parameters relevant to the models can be acquired by means of satellite imagery, in particular from the Thematic Mappers (TM), multispectral imaging sensors launched with Landsats-4, -5 and -7 in 1982, 1984 and 1999 respectively. Due to their wide spectral bands and radiometric accuracy they provide good ground resolution, sufficient to assess relevant erosion parameters (Chapter 8).

5.4.2 Aeolian erosion

With the rapid expansion of farmland, the promotion of monocultures and the removal of hedgerows to allow larger farming machinery to move, only a couple of dry years were needed until the impact of aeolian erosion became obvious to everybody (Figure 5.7). In the 1920s, dust storms were reported from the British

Figure 5.7 *Wind erosion caused by farming (Provence, France)*

Fenlands, the Brecklands, East Yorkshire and Lincolnshire (Goudie 2000). In the 1930s, dust storms were reported from Texas, Colorado, Oklahoma and Kansas. In 1992, the US National Resources Inventory stated that the annual soil loss due to wind erosion was of the order of 6 ton ha^{-1} of rural land. It is not only the American 'Dust Bowl' which is affected, however, as all over the world dust and sandstorms are degrading land and plaguing the population. In North East Asia, for instance, dust and sandstorms occur nearly five times as often as they did in the 1950s.

Wind erosion occurs when the wind speed at the ground surface exceeds the erosion threshold speed. There are three modes of movement:

- Creep: the movement of larger soil particles travelling only short distances without being detached from the ground.
- Saltation: the detachment of smaller soil particles for short time spans. When bouncing back, other soil particles are detached and thus become mobile. This transportation mode is typical for the formation of sand dunes.
- Suspended transport: the displacement of even finer soil particles over large distances. This transport mode is typical for the creation of loess deposits.

The classical work of Ralph Bagnold on the *Physics of Blown Sand and Desert Dunes* (1941) laid the foundation for understanding the basic mechanisms of aeolian erosion and transport. After many years of intensive research, wind tunnel experiments and field studies, an empirical Wind Erosion Equation (WEQ) defining the potential annual loss A_a has been derived (Woodruff and Siddoway 1965, Skidmore and Williams 1991, Fryrear *et al.* 1998, Shao 2000):

$$A_a = f(I_a, E_a, S_a, C_a)$$

where I_a is *erosivity*, the impact-triggering aeolian erosion dependent upon the storm frequency and intensity. E_a is *erodibility*, the susceptibility of the soil to aeolian erosion. S_a is a spatial component that reflects the unsheltered distance across a field. C_a expresses the influence of the soil cover and is a function of vegetation, soil profile roughness, farming and management techniques. A support practice factor is indirectly accounted for by dividing the projected year into successive management periods with characteristic wind erosion scenarios. In most publications, A_a, I_a, E_a, S_a and C_a are givens as *WE, C, I, L* and *K* (or *V*) although the apparent similarity to the RUSLE above would suggest matching parameters.

In order to model the complex interactions between these parameters the WEQ was programmed as a software tool that can be downloaded from the internet. The comparison of the WEQ approach with the RUSLE equation suggests that the processes may be combined to produce a unified soil erosion assessment tool. Attempts to achieve this are underway, for example with the *Modular Soil Erosion System* (MOSES) (Meyer *et al.* 2001).

As in the case of water erosion models, process-based wind erosion models have been developed to predict the consequences of individual storm events. The Wind Erosion Prediction System (WEPS) is such a model (Hagen 1991). Attempts to link WEPS with WEPP to derive a single process-based erosion model for water and wind erosion are underway (Fox *et al.* 2001).

5.5 Consequences

5.5.1 On-site effects

The consequences of erosion are manifold and impose on both the ecology and the economy. On-site effects include

- soil loss
- reduction of water retention capacity
- loss of filter, buffer and recharge functions
- destabilisation of natural terrain and man-made constructions.

With the loss of the upper, nutrient rich soil horizon, pasture grounds vanish and crop production decreases. Erosion rates are high in springtime when soils are usually saturated, snow is melting and vegetation cover has not yet developed. The introduction of chemical fertilisers has—in a way—aggravated the problem. Once the loss of topsoil starts reducing the productivity, fertilisers are applied without changing the farming routines. Although crop yields are kept high artificially, erosion continues and reduces the soil profile. In addition, groundwater resources are contaminated.

Erosion also damages the filter and buffer functions of the soil. Mechanical, chemical and biochemical retention and transformation processes are cut back. Contaminants are no longer filtered, nor are acids buffered. Consequently, the quality of the groundwater declines.

Another adverse effect of erosion is the destabilisation of slopes. After the vegetation has been destroyed, water can readily infiltrate into the slope. The pore water pressure increases, reducing the effective shear strength of the soil. In addition, the overall strength of the soil may decrease with increasing saturation (Chapter 4). Alpine regions that are stripped of vegetation are prone to earth and debris flows (referred to as *Muren* flows). As well as the erosion of the slope shoulders, rivers that swell after heavy rainfall may erode the toes of the slopes, thus reducing their stability. Another on-site

---

Erosion and slope failure in slums

According to the UN-Habitat Global Report on Human Settlement, 1 billion people live in slums, 32 % of today's global urban population. This number will double within the next 30 years if no concerted action is taken to alleviate the situation (UN-Habitat 2003). In slums, erosion rates remain high since building and construction work goes on continuously and uncontrollably. Streets are not paved and the remaining natural vegetation is destroyed to expand dwellings, dig for building material, harvest firewood and farm crops. Since slum dwellers are often forced up the hills surrounding the city (for example, see Figure 8), the steepness of the slopes amplifies the already considerable erosion problems. This results in excessive gullying and destabilisation leading to slope failures, especially during or immediately after heavy rains. For example, in a *favela* of Belo Horizonte (Brazil) during winter 2002/2003, a period of heavy rain caused severe gullying (Figure 5.9a) and triggered landslides claiming the lives of several people (Figure 5.9b).

effect is the destabilisation of man-made constructions. Erosion may undermine roads and railroad lines, causing immediate and expensive repair work.

According to the European Environmental Agency the overall annual on-site costs due to erosion are of the order 53 euros ha^{-1} (EEA 2003).

Figure 5.8 *Favela in Belo Horizonte, Brazil*

(a)   (b)

Figure 5.9 *Many weeks of rain during the winter of 2002–03 caused (a) gullies to form beside dwellings and (b) triggered landslides claiming the lives of several people*

5.5.2   Off-site effects

As well as on-site effects, a number of off-site effects can be identified, including:

- silting-up
- contamination
- loss of biodiversity
- flooding

Silting-up of natural watercourses (rivers, lakes) destroys aquatic ecosystems; sessile animals and plants and fish spawning grounds are covered with sediments. The turbidity of the water increases, disturbing the aquatic environment. In addition, pesticides and other pollutants spoil the water quality, reducing both biodiversity as well as the availability of clean water.

In addition, man-made structures such as canals and storm water retention ponds are silted up. Drainage networks become clogged and sewers congested. The deterioration of drainage installations and the silting up of watercourses considerably increase the risk of downstream flooding. The costs to flush and clean these installations and to excavate silted-up riverbeds and storm water retention ponds inflicts upon the budget of communities far away from the source of the problem.

Terrestrial systems are also affected by wind erosion. Farmed, natural and urban land are covered by silt and sand, either rapidly in the course of a sandstorm or continuously with dust that may originate from regions far away. This may have a detrimental effect on human health and may lead to respiratory problems, already observed in many regions regularly struck by sand and dust storms.

Eroded terrain deprived of vegetation loses the capacity to retain precipitation. In combination with the destruction of natural wetlands, the sealing of urban land and the observed increase of severe and sudden rainstorm events, disastrous floods may occur. An example is the severe flooding in Bern (Switzerland) during the summer of 2005, as depicted in Figure 5.10.

According to the European Environmental Agency the overall annual off-site costs due to erosion are in the order of 32 euro ha^{-1} (EEA 2003).

5.5.3   Climate change

It is generally agreed that carbon plays an important role in the process of global warming and thus in a frequency increase in extreme weather events (see, for example, http://www.ipcc.ch). In combination with the previously mentioned increase in flooding risk, we are facing a situation where negative effects enhance each other at a rate unseen so far. As discussed in Chapter 3, the upper soil horizon stores large amounts of CO_2 as organic carbon, about five times more than living plants (carbon sequestration). If the soil profile is deprived of its upper horizon, this function cannot be fulfilled anymore. In fact, the CO_2 storage capacity of soil even exceeds the annual CO_2 variation within the atmosphere (Kimble et al. 1998). A small reduction in the capacity of soil to bind CO_2 consequently has a strong impact on the global climate.

Obviously, the most effective way to sequester carbon is to re-establish the vegetation cover in degraded regions. If 250×10^6 ha of severely eroded land were re-greened, 0.42–0.83 Gton could be stored annually (Lal 2002). Although the sequestration rate

(a)

(b)

Figure 5.10 *The 2005 summer flood in Bern, Switzerland, (a) at the time of flooding (23rd of August) and (b) two months later when the River Aare had returned to its original level*

would peak some 15–20 years after soil restoration, restoring soil cover in combination with other means of erosion control would sequester globally some 50–70 Gton of carbon within the next 30 years, thus mitigating temporarily the greenhouse effect while allowing time to take other measures to reduce carbon emissions (Morgan 2005).

5.6 Monitoring

In the late 1980s, the United Nations recognised the need to assess the state of ground degradation on a global scale. After preliminary work of the FAO, the UNEP started a project together with the International Soil Reference and Information Centre ISRIC

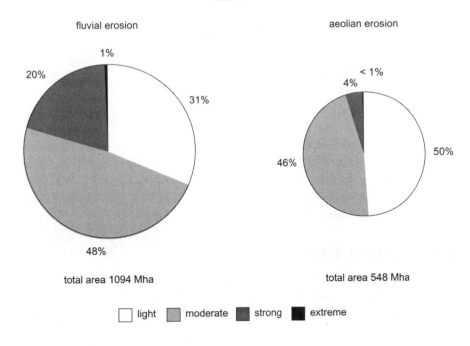

fluvial erosion   aeolian erosion

1%

20%

31%

< 1%
4%

50%

46%

48%

total area 1094 Mha   total area 548 Mha

☐ light   ▨ moderate   ▦ strong   ■ extreme

Figure 5.11 *Areas affected by water and wind erosion (after Oldeman et al. 1991)*

in Wageningen (Netherlands), aiming at a Global Assessment of the Status of Human Induced Soil Degradation (GLASOD). About 200 experts from all over the world were asked to assess the state of soil degradation in their own country against clearly defined criteria. The goal was to stimulate the awareness of the decision makers and to trigger action programmes on both a local and a global scale. A database was created, based on which the first comprehensive soil degradation map was produced. This database is continuously updated to record the global state of soil degradation and to communicate it to international organisations, decision makers and the public (ISRIC 2005).

The GLASOD data indicate that about 1.1 Gha of our planet are degraded by fluvial erosion and 550 Mha are degraded by wind erosion. Of this area, about one third is lightly, one half is moderately and the remainder is strongly and extremely eroded (Figure 5.11). Water and wind erosion comprise almost 84 % of human-induced soil degradation (Oldeman *et al.* 1991, FAO-AGL 2000). Only 16 % of a total of almost 2 Gha of degraded land is due to chemical and physical deterioration (Chapters 6 and 7). Erosion thus dominates the degradation of the Earth's soils.

6 Chemical degradation

After outlining the problem of chemical degradation, the four main aspects, notably contamination, acidification, salinization and solidification, are discussed. Basic mechanisms are explained and case files presented.

6.1 A silent spring

In 1962, the biologist Rachel Carson published her book *Silent Spring*. In this book she explains how contaminants like insecticides and pesticides had begun to poison rivers, lakes, oceans and eventually us. Chemicals have entered the food chain and we have limited knowledge about what harm they can do to the ecosystem and us and how other chemicals can enhance their effect. Rachel Carson showed that for the first time in history, humans are exposed to man-made chemicals that stay in their system from birth to death.

The chemical industry fiercely attacked her as an alarmist and tried to suppress her book, but she defended her views which were verified by the data she had gathered throughout her lifetime. In 1963 she testified before the US congress and called for new environmental policies. In a broadcasting interview for the CBS she stated that:

> "We still think in terms of conquest. We still haven't become mature enough to think of ourselves as only a tiny part of a vast and incredible universe. Man's attitude toward nature is today critically important simply because we have acquired a fateful power to alter and destroy nature. ... Now, I truly believe, that we in this generation, must come to terms with nature, and I think we're challenged as mankind has never been challenged before to prove our maturity and our mastery, not of nature, but of ourselves."

Rachel Carson died a year later, only 56 years old. With her book, she touched off an ecological awareness that eventually manifested itself in the 1992 Declaration of Rio. Al Gore credits her for laying the foundations of the US Environmental Agency (EPA), which he criticises, however, for not living up to her original goals and ideas.

Today, chemicals like DDT can still be found in the blood of polar bears, far away from the source of the pollution. Nevertheless, the chemical industry continues to push many different kinds of chemicals, promising that they will increase crop yields without harming our environment. Even worse, an unfortunate alliance has formed between the agrochemical complex and the biotechnical industry, promoting genetically modified (GM) organisms. We are now facing, for the first time in history, a type of pollution that neither dilutes nor diminishes with time but has the potential to actually multiply and mutate and adopt to changing environmental conditions. Ecologists fear that with the introduction of transgenic organisms the proverbial 'Box of Pandora' has been opened. Just as half a century ago, when Carson stood up against the chemical industry, a powerful lobby has formed to boost the contamination of our environment with organisms that haven't developed naturally: simply for profit.

6.2 Scope of the problem

Chemical degradation of natural ground has a range of aspects. The most important are:

- contamination
- acidification
- salinization
- solidification.

The first three of these four aspects are monitored globally by the GLASOD Initiative of the United Nations (Chapter 5). In addition,

- denutrification

of soil is mapped. Loss of nutrients is an indicator that soils are about to lose their natural functions (Barrow 1994). It comprises all activities that remove nutrients from the soil at a higher rate than they can be produced, including for example deforestation, over-farming and urbanisation. Denutrification weakens the resistance of the soil against erosion and may therefore be interpreted as a pre-warning sign of soil degradation.

According to the GLASOD survey, about 135.3 Mha are affected by denutrification, 76.3 Mha by salinization, 21.8 Mha by contamination and 5.7 Mha by acidification. The degree to which the land is degraded is shown in Figure 6.1. In comparison with erosion and physical degradation, chemical degradation of soil

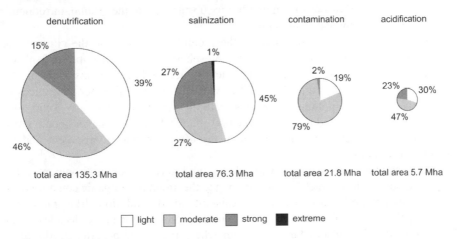

Figure 6.1 *Total areas affected by the different types of chemical degradation (after Oldeman et al. 1991)*

105

corresponds to 12.2 % of the total human-induced soil degradation (Oldeman *et al.* 1991).

The total amount of land affected by chemical degradation (239 Mha) exceeds the combined land area of Germany, France, Italy, Greece, the Iberian Peninsula and the states of former Yugoslavia. And yet, this figure tends to underestimate the dimension of the problem since only on-site effects are accounted for. Contaminants, acids and salts are mobile and enter rivers and lakes. They are transported to other parts of the world, just as Rachel Carson explained in her remarkable book that today appears as topical as it did 40 years ago.

So far, it is not clear to what extent genetically modified plants degrade soil. With the introduction of transgenic crops, landscapes deprived of biodiversity expand into ecologically intact regions. The fact that the biotechnical industry claims that transgenic plants are herbicide- and insecticide-resistant simply means that much more pesticides and insecticides will be applied. Although GM crops supposedly need less treatment than traditional crops, in practice genetically homogeneous farmland is much more vulnerable to insects and pathogens since the ability of defence mechanisms to control natural enemies is reduced. More chemicals have to be applied to suppress insects, parasites, fungus and new (unknown) diseases. This results in more pollutants being released into our environment.

In 2004, according to the International Service for the Acquisition of Agri-biotech Applications ISAAA, transgenic crops were legally planted on 81 Mha worldwide. This equals the combined area affected by salinization and acidification. In the years to come, this area will considerably increase, especially in low-income countries, which risk becoming dependent upon and exploited by a few companies marketing solutions that include GM seeds as well as chemical crop treatments with pesticides, insecticides and herbicides.

6.3 Contamination

Contamination is defined as the input of toxic substances into the ground, introducing risks to the ecosystem and to humans.

Depending on the type of pollutant, the ground conditions and the climate, the contamination propagates and damages the biosphere as well as the anthroposphere. Contaminants may disperse and subsequently accumulate, for instance in the sediments of rivers, lakes and deltas. They may also react with other substances to form more toxic compounds. Besides natural substances that are toxic, our civilisation has introduced more than 100 000 chemicals that had never been present before on this planet. For many of them, the effect on our environment and on ourselves is unknown.

Many contaminants degrade very slowly and stay in our environment for a long time, such as heavy metals which belong to the group of inorganic contaminants. Some organic pollutants are also resistant to degradation like persistent organic pollutants (POPs) that remain in our environment for many decades They have been traced at places far away from where they were actually produced or applied. In 2001, twelve of the POPs (the 'dirty dozen') were banned at the POP Convention in Stockholm, including polychlorinated biphenyls (PCBs), polychlorinated dibenzodioxines (PCDDs), polychlorinated dibenzofuranes (PCDFs) and nine pesticides.

Contamination leads to a loss of soil functions. Contaminants may be toxic to plants, animals and humans. Phytotoxic contaminants corrupt the growth of plants. Once in the soil, contaminants are taken up by terrestrial organisms and subsequently enter the food chain. Eventually they also harm humans. Carcinogenic contaminants cause cancer, fetotoxic contaminants disturb the growth of the foetus and mutagenic contaminants damage the genetic disposition. Almost all contaminants are ecotoxic, that is, detrimental to the ecosystem of which humans are an integral part.

Contamination slows down and even halts the activities of soil organisms. This weakens the resistance of the soil to erosion and ultimately destroys its buffer functions. In addition, contaminated runoff spoils surface waters and contaminated infiltration pollutes groundwater resources. The biodiversity is degraded since only contaminant-resistant organisms survive. The loss of biodiversity destabilises the vegetation cover that eventually fades, which adversely affects the climate and thus enhances the greenhouse effect.

6.3.1 Types of contaminants

Many aspects govern the behaviour and the grade of toxicity of contaminants, which makes it difficult to systematically group them into contaminant classes. The US Federal Remediation Technology Roundtable (FRTR 2006) has proposed a straightforward classification system that groups contaminants by their chemical composition and according to appropriate clean-up approaches. They distinguish between

- volatile and semivolatile organic compounds
- inorganic contaminants
- explosives
- radionuclides.

To this the group of

- agrochemicals

has to be added. Although agrochemicals may be included in the classes proposed by the FRTR, their mode of application and their ways of dissemination justifies establishing a special class of contamination.

Organic contaminants can be degraded by microbes. However, biodegradation is a complex process, difficult to accelerate and control. In certain cases, by-products more toxic than the original contaminant may be produced during the biodegradation process. Inorganic contaminants are generally not biodegradable. If they are not transformed chemically, they remain in the ecosystem until they are diluted below toxicity levels. However, under certain conditions inorganic contaminants can accumulate again, for example in lakebed and seabed sediments. Radionuclides remain dangerous to plants, animals and humans even if dispersed in small concentrations. Their duration is controlled by their half-life, which may be thousands of years.

6.3.1.1 Volatile and semi-volatile organic compounds

The large group of volatile organic compounds (VOCs) and semi-volatile organic compounds (SVOCs) can be subdivided into nonhalogenated and halogenated organic compounds, depending on whether they contain a halogene (chlorine, bromine, iodine or fluorine). Nonhalogenated compounds include

- *Nonhalogenated VOCs*, such as paints, thinners and solvents used for dry cleaning and metal degreasing and *nonhalogenated SVOCs* having a smaller vapour pressure, making them less volatile.
- *Polycyclic aromatic hydrocarbons* (*PAH*): chemical compounds that contain more than one fused benzene ring, of which traffic emissions are a major source.
- *Fuels*, for example gasoline, diesel and kerosene. They can usually be found close to filling stations, at aircraft and vehicle maintenance areas and sites of oil spills.

Halogenated organic compounds tend to be more resistant to biodegradation than non-halogenated compounds. They thus pose a higher danger once released into the environment. Typical halogenated compounds are

- *Polychlorinated biphenyls* (*PCBs*), which are found, for example, in the oil of electrical transformers. They were banned at the Stockholm POP-convention 2001.
- *Pentachlorophenols* (*PCPs*), which are typical for wood-preserving sites.

VOCs and SVOCs may be found at sites of production, utilisation, application and handling of chemicals such as metal finishing shops, paint stripping sites, spray booth areas, degreasing sites and old gas works, as well as at disposal sites, burn pits, leach fields, landfills, maintenance areas, along traffic and railroad installations, close to storage tanks and pipelines and, of course, at sites of chemical compound production.

6.3.1.2 Inorganic contaminants

Inorganic contaminants cannot biodegrade and thus remain in our environment until they are diluted or chemically altered. According to the FRTR (2006) they are grouped into

- Metals including barium, cadmium, chromium, copper, lead, mercury, selenium and zinc. Arsenic is also included in this group although it is a metalloid (a semimetal) rather than a true metal.
- Other inorganic compounds such as asbestos, cyanide and fluorine.

Inorganic contaminants are found at chemical plants and mining sites, at sites of utilisation, application and handling of inorganic compounds such as metal finishing shops, paint stripping sites and battery refilling shops, at storage and disposal sites, at leach fields, at military bases and training areas and at derelict steel works.

6.3.1.3 Agrochemicals

A special group of contaminants are agrochemicals. They are typically introduced in a diffuse way on farmland to enhance crop yields. Since they are regularly over-applied, they leave the farmland as runoff or infiltrate the groundwater. Two special groups of agrochemicals are discussed:

- pesticides and
- fertilisers.

Pesticides are chemicals that are applied to prevent, destroy, repel or mitigate any pest. They include insecticides, fungicides, herbicides, acaricides, nematodicides and rodenticides. Pesticides are SVOCs that may be either halogenated or nonhalogenated. DDT is a typical pesticide that was applied to kill insects until the early 1970s. Many countries recognised that DDT is also dangerous to the ecosystem and that it threatens human health. Eventually, it was banned at the Stockholm POP Convention together with aldrin, chlordane, dieldrin, endrin, heptachlor, hexachlorobenzene, mirex and toxaphene, all being persistent pesticides. However, most of these pesticides need decades to decompose and can still be found at locations around the world, from the rainforest to the polar caps. Figure 6.2, for example, highlights how the concentration of pesticides DDT and lindane within Swedish soils, untreated with these chemicals, increases with latitude. Figure 6.3 demonstrates that some of the pesticides used resemble agents that have been employed in chemical warfare.

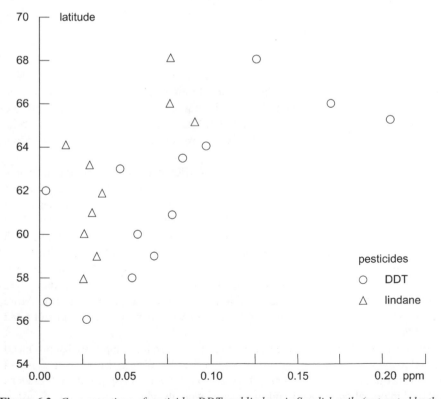

Figure 6.2 *Concentrations of pesticides DDT and lindane in Swedish soils (untreated by these chemicals) in the 1960s (after Oden 1970)*

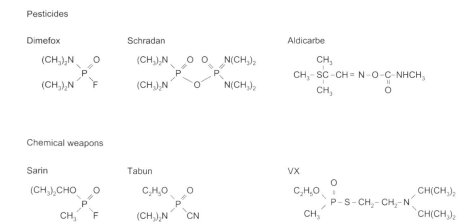

Figure 6.3 *Comparison of the molecular structure of certain insecticides with chemical weapons (after Ramade 2005)*

Operation Ranch Hand

From 1961 to 1971, *Agent Orange* was applied during the Vietnam War as defoliant to clear rainforests and mangroves. The aim was to remove the vegetation that was sheltering the Viet Cong and to destroy their food resources. Agent Orange is one of the 'Rainbow Herbicides' that included also Agent White, Green, Blue, Pink and Purple, referring to the stripes painted on the barrels containing the defoliant. It consists of a mixture of two different herbicides that were developed in the 1940s and contains 2,3,7,8-tetrachlorodibenzo-para-dioxin (TCDD), one of the most toxic members of the dioxin family. TCDD is a human carcinogen that causes even in a very small concentration chloracne and many forms of cancer including leukaemia, respiratory and prostrate cancer. It is also believed to cause spontaneous abortion, stillbirth and severe birth defects.

During the operation 'Ranch Hand' about 40 000 cubic metres of Agent Orange were sprayed over Vietnam, Cambodia and Laos. More than 300 kg of TCDD was released over southern Viet Nam alone (Stellman *et al.* 2003), where thousands of square kilometres are still contaminated. The association *Vietnam, les Enfants de la Dioxine* (Ivry, France) and the *Collectif Vietnam Dioxine* (Paris) claim that 2 to 5 million Vietnamese have been exposed to Agent Orange including many children. More people have been affected in Cambodia and Laos. Since TCDD has spoiled the soil, it has also entered the food chain and thus continues to pollute the daily diet, 40 years after it was applied. Today, children are born with grave birth defects. More than ten *Peace Villages* have been established in Viet Nam in order to aid and tread the victims.

In 2005, a lawsuit filed by Vietnamese victims against the companies that produced the defoliant was dismissed by the Brooklyn Federal Court, reasoning that Agent Orange was not considered a chemical weapon at the time of application. US citizens involved in operation 'Ranch Hand' have, however, already been compensated, just as their allies from Australia, Canada, New Zealand and South Korea.

In lieu of organic manure, chemical fertilisers have been applied worldwide from the middle of the 19th century with the aim of raising fast crop yields. Chemical fertilisers are intended to replace the nutrients extracted from natural soil by crop plants, removed during harvest. The global application of chemical fertilisers increased by a factor of 18 from 1946 to 1988, when 146 Mton were applied (Ramade 2005). Although the utilisation of chemical fertilisers has slowed in wealthy countries, their application is still increasing in low-income countries. Figure 6.4 demonstrates that since 1989, the amount of fertilisers applied has increased globally by 15 percent while the amount of pesticides consumed has remained on a high level, indicating the importance of chemicals in agricultural production. The number of animals per hectare of permanent pasture has increased whereas the number of workers per hectare has decreased, indicating a further intensification and mechanisation of agriculture. On the other hand, the share of agricultural GDP of total GDP has fallen.

Nitrogenous, phosphate and ammonium-rich fertilisers are converted to nitrate compounds by bacteria, rendering rivers and lakes eutrophic, contaminating the groundwater and enhancing the greenhouse effect with nitrogenous gases (Barrow 1994). In addition, chemical fertilisers and other impurities enter the soil. Superphosphate fertilisers carry metals and metal-like compounds including cadmium, chrome, cobalt, copper, lead, nickel, selenium, vanadium, zinc and arsenic. The utilisation of sewage sludge as fertiliser introduces an additional source of contamination.

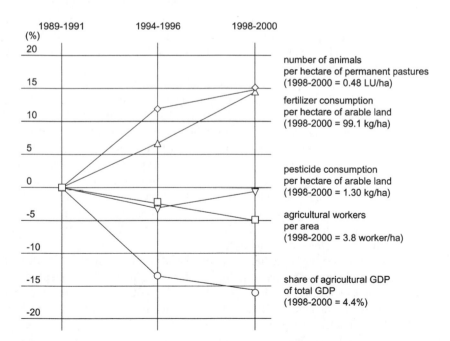

Figure 6.4 *Changes in agricultural practices (percent) during the decade from 1990 to 2000 (from FAO world statistics 2005)*

6.3.1.4 Explosives

The group of explosives comprises (FRTR 2006)

- Propellants, including rocket and gun propellants.
- Explosives, including primary explosives such as lead azide and lead styphnate that are frequently used to ignite secondary explosives like melt-poor explosives based on trinitrotoluene (TNT), plastic-bonded explosives (PBX) and crystalline explosives such as RDX (cyclotrimethylenetrinitramine or cyclonite).
- Pyrotechnics, including flares, smoke generators, incendiary delays, fuses and photoflash compounds.

Propellants, explosives and pyrotechnics are summarised as PEPs. Typically, PEPs can be found on military bases, shooting ranges and disposal sites as well as in war-affected zones.

6.3.1.5 Radionuclides

Radionuclides are radioactive elements characterised by their carcinogenic and mutagenic effects on living organisms including humans. Since they usually have a very long half-life, they stay quasi-permanently in our environment. Radionuclides may originate from sites of uranium mining and processing. Uranium mining in eastern Germany has left more than $300 \times 10^6 \, m^3$ of spoil heaps and more than $160 \times 10^6 \, m^3$ of radioactive sludge in densely populated regions. After the German reunification, the scale of the problem was recognised and remediation measures began, focussing on removing the spoil heaps and covering the sludge lakes. The costs of the remediation measures are estimated to total almost 6–7 billion euros.

Radionuclides can also be found in hospital and laboratory wastes, as well as in wastes from industries processing and handling radioactive material. It can also happen occasionally that radionuclides are accidentally released. The Chernobyl disaster demonstrated that the migration of radionuclides can hardly be controlled and regions beyond the borders of Russia have been affected by the nuclear fall-out. In addition, the nuclear waste produced is difficult and expensive to store and research is still ongoing to find practicable and cost-efficient solutions. In the United States, the detritus of the nuclear age includes some 50 000 ton spent fuel from commercial, military and research reactors, as well as 350 ML of radioactive waste from plutonium processing. Most of this waste will likely go to the 50-billion-dollar repository at the Yucca Mountains in Nevada (Long 2002).

As well as this, testing of nuclear weapons has degraded vast areas in the United States, Russia and many other parts in the world. According to the US Department of Energy, the Nevada testing site alone extends over 360 000 ha, larger than the state of Rhode Island. Until the Nuclear Weapons Moratorium of 1992, it has seen more than four decades of nuclear weapons testing. Today, it is still used to test conventional weapons, monitor hazardous chemical spills and train emergency response teams. In Russia, the Arctic island of Novaya Zemlya has been the theatre of nuclear tests since the beginning of the Cold War. More than half of the almost 10 Mha of the archipelago is officially commissioned as a testing range. The indigenous people that used to live on Novaya Zemlya have been transferred to other parts of Russia.

Moreover, the use of depleted uranium (DU) to increase the weight of conventional shells and grenades has lead to a widespread pollution of combat zones with radionuclides, although nuclear weapons were never applied. This was first proven in a UN document on the Kosovo War (UNEP/UNCHS 1999). DU is a waste product from enriching uranium ore for nuclear reactors. Since DU is extremely dense, it is used on the tip of bullets to ease their penetration into armoured vehicles.

To summarise, there are many sources of radioactive contamination, but efficient approaches to dealing with them are difficult to develop.

6.3.2 Fate of contaminants

The fate of contaminants is controlled by their physical and chemical properties, the way in which they are released into our environment and the medium into which they are deposited. With regard to the mode of release we distinguish between

- point sources such as spills and accidents, broken pipelines and rotting barrels, illegal dumping sites and engineered landfills, as well as
- diffuse sources such as agrochemicals that are disseminated over farmland or traffic exhausts that migrate with the atmosphere.

As long as contaminants are mobile, they may migrate in one of several ways. Atmospheric transport involves toxic gases, liquids and dust particles. The pollutants are diluted and dispersed in the air and carried with the wind to locations that may be far from their place of origin. This transport mode is driven by pressure gradients and is referred to as advection. If there is no wind to move the contaminants, they may still be displaced due to a concentration gradient. This transport mode is called diffusion. Eventually, contaminants are deposited as dust or fall-out as rain. With that, a diffuse contamination is introduced that may affect whole regions.

Fluvial transport refers to the movement of contaminants with the flow of water. Dissolved contaminants travel as aqueous phase by advection as well as by diffusion due to concentration differences. Insoluble contaminants travel as non-aqueous phase. Depending on their specific weight and the velocity and type of flow (laminar, turbulent) they may be dispersed, sink to the bottom or float on the surface. Soil may be affected by fluvial contamination when exposed to contaminated runoff or storm water.

Once contaminants enter the soil, they occur in one of four phases:

- gaseous phase in the unsaturated zone
- solid phase adsorbed to soil particles
- aqueous phase dissolved in pore water and groundwater
- immiscible phase as non-aqueous phase liquids (NAPLs).

The migration pattern depends on the type of the ground, notably its permeability (Chapter 4), as well as the type of the contaminant and its biodegradability. The upper active layers of a soil profile retain considerable amounts of contaminants due to their buffer functions (Chapter 3). As soon as the retention capacity of the solum is exhausted, contaminants enter the parent material that may either be a sedimentary deposit or bedrock. Fine sediments like silt and clay have a low permeability and consequently retard and immobilise contaminants. Coarse sediments like sand and gravel have a high permeability and thus allow rapid migration. In fractured bedrock, contaminants migrate along the fractures. A portion of the contaminants is adsorbed at the fracture walls and a quantity actually enters the intact rock, an effect typical for porous rock like sandstone. This migration pattern may be relevant for sediments as well, since they may also be fractured (Chapter 4). Transport equations distinguish between solute and phase transport as well as porous and fractured media. Since natural ground is usually inhomogeneous and anisotropic, modelling transport problems is a demanding task.

VOCs and SVOCs may adopt all four forms of migration. As pore gas, VOCs migrate independently of the groundwater flow by advection (pressure gradient) and diffusion (concentration gradient). Dissolved in the groundwater, organic compounds travel by advection and diffusion. Light non-aqueous phase liquids (LNAPLs) float on the groundwater table as shown in Figure 6.5 (for example, diesel fuel or crude oil) whereas dense non-aqueous phase liquids (DNAPLs) (chlorobenzene, naphthalene or coal tar) sink to the base of the aquifer (Figure 6.6).

Metals accumulate in the solum until the metal retention capacity is exhausted. They may be solubilised due to the acidification of the soil. In this case, metals start migrating with infiltrating rainwater and contaminate the groundwater. In West Bengal and Bangladesh, about 30 million people are exposed to arsenic soil and groundwater. The use of pesticides, industrial wastes as well as coal and wood combustion has increased the level of arsenic, which is also naturally present in the

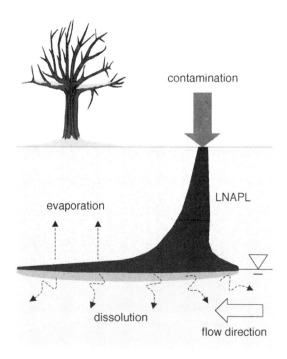

Figure 6.5 *Migration pattern of LNAPLs*

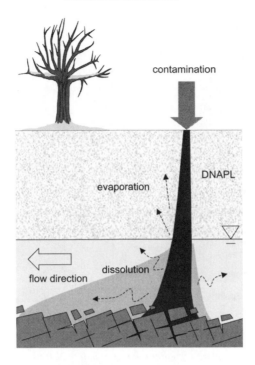

contamination

DNAPL

evaporation

flow direction

dissolution

Figure 6.6 *Migration pattern of DNAPLs*

sediments of the region. Arsenic is colourless, odourless and tasteless and so is readily taken up by animals and humans, causing cardiovascular problems, skin diseases, peripheral neuropathy and kidney damage (UN-Earthwatch 2005).

With regard to possible remediation options (Chapter 9) we may distinguish between local contamination and diffuse contamination. Local contamination is characterised by hot spots that have to be extracted or neutralised in order to avoid spreading of the pollution. In general, local contamination is associated with urban and suburban environments. However, a local contamination can also be found away from the cities, for instance at military bases and mining sites. Diffuse contamination is much more difficult to handle since it is not confined to a limited area that can readily be treated. As will be shown later (Chapter 9), remediating a diffuse contamination calls for special treatment approaches and sometimes involves fundamental changes in land management practices.

In Europe, diffuse contamination with agrochemicals prevails in the Western Lowlands and in the eastern accession states. Local contaminations are usually associated with industrial regions and are thus typical for the Ruhr District, the British industrial belt, Pas-de-Calais in France, northern Italy, the region around Cracow and Katowice, and many other places of industrial activity. Many countries spend substantial funds cleaning up these sites (Figure 6.7). Industrial contamination being washed into storm water drains and then rivers is a major problem in all industrialised nations. In the East, military bases also pose a major problem, especially in the Baltic States, the Czech Republic and in Hungary (EEA 2003).

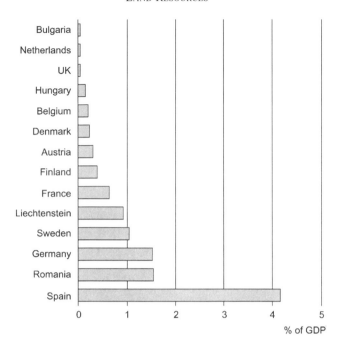

Figure 6.7 *Expenditure on the remediation of contaminated sites by certain countries in 1999 as a percentage of GDP (after EEA 2003)*

The Pintsch Site (Genske 2003)

The Pintsch Site in Germany is a typical example of local contamination. In 1942, oilcontaminated soil was dumped adjacent to the River Main in the city centre of Hanau. Later, the Pintsch Company was established on that site to treat and recycle waste oil arriving from all over Germany. Careless handling of the products, however, led to a severe hydrocarbon contamination of the ground. In 1984, shortly after the environmental damage was discovered, the company was closed down by municipal order. At that time, it was one of the largest reported in Germany.

The ground is characterised by an upper layer of fill (1–2 m thickness), followed by loam (0.5–1.8 m) and Quaternary terraces of the River Main (5–8 m). The Quaternary is underlain by a 50-m thick layer of fine Tertiary sediments, which are in places interlocked with basalt layers of several metres thickness. The groundwater table is located in the Quaternary at a depth of about 3.5 m, and fluctuates with the water level of the River Main. An oil phase of up to 70 cm thickness was detected floating on the groundwater table (Figure 6.8). Furthermore, hydrocarbons were detected in the unsaturated zone and in the aquifer itself. In order to recover some 400 tons of oil floating on the groundwater table, recovery wells were installed creating depression cones in which the LNAPLs accumulated. With vertical barriers keyed into the fine-grained Tertiary sediments, the total amount of contaminated groundwater to be pumped could be reduced. It is estimated that 30–40 % of the floating oil phase was finally recovered after several years of pumping. Microbiological *in situ* methods were applied to degrade the remaining contaminants.

The Chernobyl Disaster

The Chernobyl Disaster is an example of regional contamination. The greatest catastrophe involving a nuclear power station to date occurred in Pripyat, Ukaine, during the night of April 25–26th 1986. Officially, in Belarus, Russia and the Ukraine 14 Mha are still considered contaminated by Caesium-137 at an intensity of over 37 000 Bq m^{-1} (1 Bq = 1 Bequerel). Caesium-137 has a half-life of about 30 years. About half of the total quantity of Chernobyl's volatile inventory was deposited outside these three countries, and nuclear fallout was recorded in Scandinavia, the United Kingdom, Germany, France and even in North America and Japan. Restrictions on food contaminated due to the fallout are still in place, as for instance in the United Kingdom, Sweden, Finland, Lithuania, Poland, Germany, Austria and Italy (Fairlie and Summer 2006). Regardless, about 200 000 people reside on contaminated terrain although hundreds of thousand of people have already been evacuated from the most severely affected areas. At the Closed Zone, the radiation exceeds 1 480 000 Bq m^{-1} (see Figure 6.9 for map of the zone limits).

In 1992, the Ukraine spent 15 % of the state budget to remediate the effects of the Chernobyl disaster. In 2015, the economic damage will have accumulated to about 200 billion dollars in the Ukraine alone (Fairlie and Summer 2006). According to Angelina Nyagu, president of the Ukrainian foundation *Physicians of Chernobyl*, more than seven million people still suffer today from the effects of the catastrophe. Illnesses include cancer, cataract induction and cardiovascular diseases, heritable effects (damages to genes and chromosomes) as well as mental and psychosocial effects.

This contradicts an IAEC press release in September 2005, which summarised IAEC and WHO reports which stated that "a total of up to 4000 people could eventually die of radiation exposure from the Chernobyl nuclear power plant accident".

6.4 Acidification

Acidification is the process which renders soils acidic. According to the FAO, strongly acidic soils are characterised by a pH below 5.5 and extremely acidic soils by a pII below 4.5. Acidic soils have a typically low cation-exchange capacity and a low base saturation.

Acidification of soils results in a loss of soil functions. On natural land, the flora and fauna is damaged, sometimes to an extent that particular plants perish and certain animals retreat. A widespread forest die-back syndrome, the *Waldsterben*, was first reported in the late 1960s from the Black Forest, Germany. Soil acidification alters the metabolism of plants, hinders photosynthesis and disturbs microbial activities in the rhizosphere. As a consequence, biodiversity is reduced both within the soil and on the surface (Figure 6.10). On farmed land crop yields drop and remediation measures become necessary to restore soil fertility. In addition, acid runoff affects rivers and lakes and acid infiltration lowers the pH of the groundwater and may thus mobilise metals.

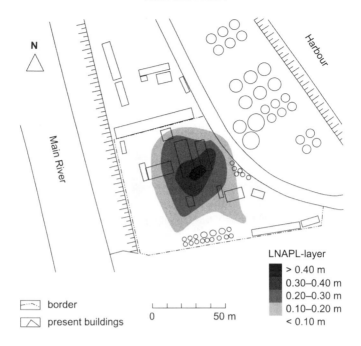

LNAPL-layer

■ > 0.40 m
■ 0.30–0.40 m
■ 0.20–0.30 m
■ 0.10–0.20 m
< 0.10 m

border
present buildings

0   50 m

Figure 6.8 *Map of the terrain at initial detection, depicting the floating oil phase (after Ripper 1992)*

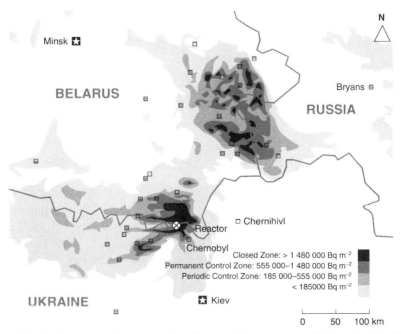

Closed Zone: > 1 480 000 Bq m⁻²
Permanent Control Zone: 555 000–1 480 000 Bq m⁻²
Periodic Control Zone: 185 000–555 000 Bq m⁻²
< 185000 Bq m⁻²

0   50   100 km

Figure 6.9 *Radiation resulting from the Chernobyl Disaster, simplified (after Handbook of International Economic Statistics 1996)*

118

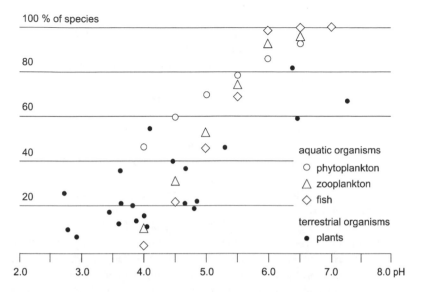

Figure 6.10 *The reduction of biodiversity within the soil and on the surface due to the acidity of the rain (after Keller and Gunn 1995, Bradshaw and Chadwick 1980)*

6.4.1 Natural acidification

Acid soils occur naturally in the tropics and subtropics, but also in moderate climates, especially in boreal regions. High precipitation rates prompt acidification, since soil nutrients such as calcium, magnesium, potassium and sodium are leached below the root zone while being replaced with hydrogen. This process is further stimulated by weak acids that are produced during the decay of the surface litter, that is, the decomposition of plant residues and organic matter.

Podzolization is the natural process of leaching soluble minerals from the A-horizon to the B-horizon in a low pH-environment and is typical for non-alkaline ground in cool and humid climates. Acrisols and ferralsols are often classified as acidic and arenosols, cambisols, histosols, luvisols, planosols and fluvisols tend to become acidic.

6.4.2 Human-induced acidification

Many factors may lead to the acidification of soil. The main causes are

- unsustainable farming practices
- deforestation
- acid rain producing emissions.

The conversion of natural land into farmland initialises an acidification of the soil (Goudie 2000). This may have already begun with slash-and-burn clearances and involves complex biochemical processes. Overfarming may remove enough bases to acidify soils, as does the application of certain agrochemicals. In particular, the use of nitrogen fertilisers has been proven to acidify farmland. Soil bacteria nitrify the ammonium nitrogen introduced with the fertiliser, while releasing free hydrogen ions that render the soil acid. If fertilisers are applied excessively, the nitrate not taken up by the plant is leached below the root zone together with other nutrients such as calcium and magnesium, which again acidifies the soil. A low pH is accompanied with high aluminium saturation. Al-toxicity causes a decline in crop yields since root growth is hampered and the rate of nutrient uptake is reduced.

In 1852, Robert Angus Smith started publishing on the chemistry of rainwater around the city of Manchester, UK, and summarised his work 20 years later in his book *Air and Rain: The Beginnings of a Chemical Climatology*. In this book based on data from England, Scotland and Germany, he proved that acid rain is caused by the combustion of coal. He also described the damage incurred on plants and soil. His book, however, remained unnoticed, just as later works by Eville Gorham on acid rain and soil degradation in the United Kingdom and Canada. The Swedish soil scientist Svante Oden brought the issue to a broader audience when he toured the United States and reported at international conferences in the 1970s. Sweden intervened at the 1972 Stockholm Conference on the Human Environment and concerns were voiced about the fact that the lakes and soils of Scandinavia were acidifying due to emissions in countries south and east of Sweden. Figure 6.11 illustrates how the acidity of rainfall over Sweden increased until the 1980s, which in turn would have led to a decline in species number for both terrestrial and aquatic ecosystems. The First International Symposium on Acid Precipitation and the Forest Ecosystem was eventually held in Ohio in 1975 and from that point onwards, the scientific community began to acknowledge the dimension of the problem (Cowling 1982).

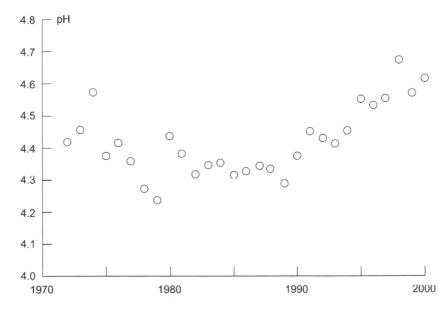

Figure 6.11 *The acidity of rainfall over Sweden increased during the 1980s (after Swedish Environmental Protection Agency)*

Barrow (1994) proposes the use of the phrase acid deposition instead of acid rain, since the acidification of soil may be wet (rain, mist, droplets) as well as dry (dust, gas, smog). The major cause of acid deposition is the emission of sulphur dioxide. Although there are many natural emission sources (e.g. volcanoes, underground coal and oil shale fires, bush fires and microbiological degradation processes), it is due to the combustion of coal to produce energy, ore smeltering and other man-made sources that the total output of SO_2 doubled by the end of the 1970s. Other man-made emissions, for example nitric oxides (NO) and nitrous oxide (N_2O), react with SO_2 and increase its impact.

Shallow soils are much more vulnerable to acidification than deeper soils because of their reduced buffer capacity. Soils on slow weathering, non-alkaline bedrock such as granite and soils that develop on sand, gravel or glacial deposits are much more affected by acid deposition than alkaline soils that develop, for example, on limestone. The young and shallow Scandinavian soils that developed on the Precambrian bedrock after the glaciers of the ice age retreated are particularly susceptible to acidification, as are a large portion of Canadian soils.

In Europe, soil acidification from acid rain has decreased by 50 % since the 1980s owing to a reduction in sulphur emissions. Where it persists, however, it continues to degrade the land (UNEP 2002). It is expected that in Central and Eastern Europe soil acidification will continue to some extent.

The acidification of Sudbury

A well-documented example of the degrading effects of acid deposition is the mining district around Sudbury, Canada (Gunn 1995). Sudbury used to be one of the biggest mining locations for metals in the world, mainly nickel and copper. While mining commenced, ore refinement installations were established. In the early days, open-air roast yards (for example, Figure 6.12) and smelters were introduced to process Sudbury's sulphide ore, for which a tremendous amount of timber was needed. The trees not cut for roasting and smelting later perished due to an alarming amount of air pollutants that fell out as acid and toxic rain. The combination of clear-cutting, erosion and air pollution virtually wiped out all vegetation leaving behind what was locally referred to as 'moonscape' (see Figure 6.13 for a map of the degraded region). Even the outcropping rock was altered chemically into a dark-coloured crust (Figure 6.14). Sudbury became one of the largest point sources of air contaminants with emissions comparable to the entire sulphur dioxide release of the United Kingdom in the 1990s. 17 000 ha of land were industrially degraded and 7000 lakes became acid (Genske 2003).

6.5 Salinization

Salinization is a process that increases the salt content in the soil and results in a loss of soil functions. Soil fertility is greatly reduced, in many cases to an extent that farmland is abandoned. In the United States, salinization reduces crop yields in one-quarter to one-third of the irrigated land. In Mexico, salinization causes a loss of 1 Mton of grain every year, the equivalent of enough food to feed 1 million people (Stockle 2001).

Figure 6.12 *The O'Donnell roast yard before ignition in 1920 (photograph courtesy of W. McIlveen/Inco Archives)*

In the same way that farmland is affected by salinization, so too is natural land. As the salt level rises, the osmotic potential of the roots degrades, making it more difficult for the plant to extract water. At a certain salt level the plant may actually dehydrate. In addition, high salt levels may also lead to a change in soil structure towards more compacted soil conditions, especially in clay-rich soils. With decreasing pore space the

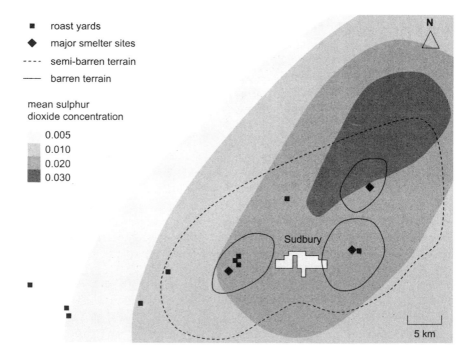

Figure 6.13 *The Sudbury Basin with the major sites of roasting and smeltering and mean sulphur dioxide concentrations, used to categorize fumigation zones (1953–68) and degree of land degradation in 1973 (after Winterhalder 1995)*

122

Figure 6.14 *Even the rock surface became dark because of acid rain*

conductivity of water and air decreases as well, which hampers the development of the plant. Eventually, the plant dies and the soil becomes vulnerable to erosion. Barren land develops and biodiversity vanishes. In addition, saline runoff spoils rivers and lakes and saline infiltration damages the groundwater resources.

6.5.1 Natural salinization

Salt affected soils occur naturally and are characterised by considerable amounts of soluble salts. They can be grouped into

- saline soils and
- sodic soils.

Saline soils are typical for the dry tropics and subtropics (deserts). According to the FAO, saline soils are characterised by an electrical conductivity of more than 15 decisiemens per metre. The reference soil is the *solonchak* with a salic (salt rich) horizon, which develops when evaporation exceeds precipitation. Salt rich solutions rise to the B-horizon where they fall out or even to the surface where they form salt crusts. The salt present in the soil solution may originate from the parent material, from a saline groundwater table or from infiltrating salt rich water. Salic horizons may also be found in histosols, vertisols and fluvisols.

Sodic soils occur typically in the dry midlatitude zone (steppe). The reference soil is the *solonetz*, which is characterised by a humus rich surface referred to as a natric-horizon (rich in clay, natrium and sodium) and saline subsoil that is recognisable by a typical columnar pattern. Solonetz soils are strongly alkaline with a pH of more than 8.5. They may be associated with histosols, gleysols, chernozems, kastanozems, vertisols and solonchaks (FAO-AGL 2005).

6.5.2 Human-induced salinization

The salt content in soil may be increased artificially because of improper soil management routines. Human-induced salinization has a variety of causes including:

- unsustainable farming practices

- deforestation
- groundwater table ascent
- saltwater intrusion
- saline runoff and infiltration.

When farmland is irrigated, water evaporates and salt concentrations increase. As early as 6000 years ago, the early cultivations of the Mesopotamian civilisation diverted water from the Euphrates River to irrigate their fields. The following Sumerians founded the first irrigation-based civilisation that collapsed after 2000 years of good harvests, as their fields turned into a salty wasteland (Stockle 2001).

It is estimated that irrigating land in a dry climate with $10\,000\ \mathrm{m^3\,ha^{-1}}$ of water during the growing season introduces an average of $5\ \mathrm{ton\ ha^{-1}\,yr^{-1}}$ of salt (Ramade 2005). When brackish water is used for irrigation, the process of salinization rapidly accelerates. In addition, the application of high salt index fertilisers considerably increases the salt content of both soil and runoff.

Moderate to high salinization is already affecting agricultural soils in the Mediterranean region and the countries of Eastern Europe, the Caucasus and Central Asia (EECCA), mainly as a result of inappropriate irrigation systems (see Figure 6.15). For example, salinization affects 16 Mha or 25 % of irrigated cropland in the Mediterranean (EEA 2003) and about 30 % of all irrigated cropland worldwide (Ramade 2005). Some 10 Mha of irrigated land are destroyed due to salinization every year (Pimentel *et al.* 2004).

In areas where land subsides, due to the pumping of oil for example, the distance to the groundwater table may decrease in such a way that capillary forces transport groundwater enriched with soil minerals to the surface where it evaporates, leaving salt patches that may combine to form a large salt layer. In certain regions with long

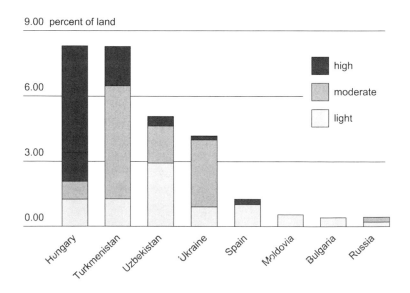

Figure 6.15 *Land affected by salinization in particular European countries (after EEA 2003)*

and intensive rainfall clear-cutting of the forest may cause the groundwater table to rise, thus prompting salination problems.

Over-pumping of the groundwater close to the coastline leads to the intrusion of saltwater, which may render soil saline as in Israel for example and many countries of Arabia and even the United States. However, measures taken far inland may also lead to saltwater intrusions. For example, the construction of the Aswan Dam (Egypt) caused groundwater levels to drop downstream, prompting seawater to salinate soils along the delta (Goodie 2000).

The mining industry has contributed considerably to the salination of vast regions. Salt-bearing refuse heaps are scattered around salt mines (see, for example, Figure 6.16) and contaminate surface and groundwater as well as the adjacent soils. Close to cities, the infiltration of municipal wastewater increases the salt content, as does the application of de-icing salt on roads and sidewalks.

The Aral Sea Disaster

From the 1950s onwards, water has been diverted from the two major rivers feeding the Aral Sea—the Amu Darya and the Syr Darya—to irrigate cotton fields. This irrigation programme, orchestrated by the former Soviet Republic, has created a threefold salinization problem.

Firstly, poor irrigation practice has led to widespread salinization and waterlogging of the new farmland. The low irrigation efficiency is due to the small percentage of lined canals and the inadequate state of the drainage network that is poorly maintained. After only 40 years, about half of the irrigated land has already become saline. Secondly, the diminishing water of the Aral Sea as well as the groundwater has become saline. From 1960–65 i.e. within only 5 years, the level of the Aral Sea fell by 19 m, the surface area shrunk by more than half (see Figure 6.17 for changing shoreline) and the salinity of the lake water trebled. Today, only three highly saline lake fragments remain of what was once the world's fourth largest lake. Moreover, the salt content in the groundwater has risen to up to $30 \, g \, l^{-1}$. Finally, the dried-out lakebed has turned into a hostile salt desert, with soils highly contaminated by agrochemicals from the cotton industry and other pollutants. Since the 1970s, dust storms spread tens of millions of tons of contaminated dust and salts up to 1000 km every year.

In the meantime, the international community has recognised the Aral Sea basin as one of the major ecological disaster zones of the world. In 1993, all five Aral Sea basin countries (Kazakhstan, Kyrgyzstan, Tajikistan, Turkmenistan and Uzbekistan) signed an agreement to remediate the environmental damage but so far, due to the political instability of the region, little progress has been made (FAO-Aquastat 2005, Kobori and Glantz 1998).

6.6 Solidification

Certain soils have a tendency to develop solid layers that are also referred to as hard pans. The development of hard pans may be driven by natural processes, as in tropical plinthosols for example, where leaching of parent material and re-deposition of

Figure 6.16 *Refuse from salt mining in Thuringia, Germany*

minerals (especially iron oxides and aluminium oxides) leads to the formation of a solid layer of sometimes considerable thickness, a process also referred to as *lateritization* (*later* is Latin for brick). Alluminium-rich laterites are also called bauxite and are mined to produce aluminium. When exposed to the air, these layers may dry out quickly and irreversibly and develop a massive crust, referred to as an ironstone hardpan.

Deforestation and over-farming make tropical soils vulnerable to lateritization. Once the protective cover is eroded, the plinthic horizon becomes exposed and solidifies. The indurated laterite layer may be used as building material. The ancient

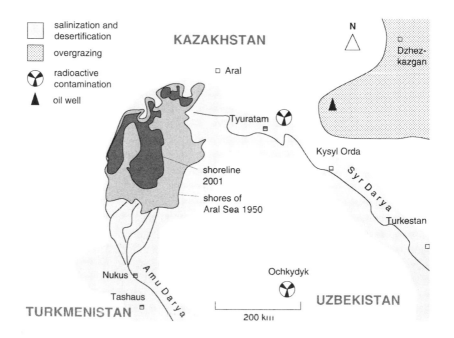

Figure 6.17 *Salinization and desertification in the Aral Basin, simplified (after UNEP-GRID 2004, FAO-Aquastat 2005)*

126

Khmer constructed their temples with this material and in India the hardpan is still referred to as 'brickstone'. However, terrain affected by lateritization loses its fertility and thus cannot be farmed anymore. In addition, the hardpan seals the ground and therefore blocks groundwater recharge. The ecological functions of the soil are destroyed. In addition, any construction of buildings and roads proves difficult on plinthic hardpans.

Although the process of lateritization was classed as 'pedological leprosy' as early as half a century ago, it hasn't yet been stopped because of mismanagement and profit-oriented exploitation of land. Lateritization is thus a good example of the deterioration of soil as a result of human-induced acceleration of pedogenic processes.

7 Physical degradation

The scope of the problem is outlined and the types of physical degradation, notably compaction, sealing and overbuilding, subsidence and waterlogging are introduced. As in the preceding chapters, basic mechanisms are explained and case files are presented.

7.1 Notions of a post-industrial era

"As time went by and things got worse, people had to move from the cold mill to the hot mill, or vice versa, just to have work. LTV Steel had purchased J & L, and they started whittling away at its employees. First to close at Eliza were the blast furnaces, and then the hot mill. I was there for the last rolling. As the coils were processed through the cold mill, I watched each succeeding unit go down. Boy, when that big, old hot mill shut down and the last red-hot bar went through, we all knew it was the end. The up-river crane man started blowing his siren, and then the center crane man started his siren, and the next crane man started his siren, until all of the sirens were blowing throughout the whole mill. It was a roaring madhouse. It was heartbreaking—big husky steelworkers had tears in their eyes. It was the end of an era."

Bob 'Ike' Eisengart describes the last day of Eliza Steel Mill in Pittsburgh in 1979 (Lane and Perrott 1989). Many impressive structures were erected during the times of industrialisation, all over the world. They consumed a large amount of natural land and destroyed precious soils. Now, we have reached the post-industrial era, and many installations that were once so important for the wealth of the nations have been abandoned. Derelict industrial terrain still reminds us of the past days of industrialisation and sky-rocketing profits. Today, globalisation forces markets to re-orientate and strive for new chances. While old industries are abandoned (see, for example, Figure 7.1), new complexes are built at ever-increasing speeds. More land is developed to accommodate workers, administrators and services. More and more terrain is overbuilt, sealed and compacted.

7.2 Scope of the problem

Physical degradation of natural ground can occur in different ways, most importantly

- compaction
- sealing and overbuilding
- subsidence and
- waterlogging.

Figure 7.1 *The abandoned Steel Mill Duisburg-Nord, Germany, now a landscape park*

Of these four aspects, compaction, subsidence and waterlogging have been monitored globally by the GLASOD Initiative of the United Nations (Chapter 5). According to their survey, about 68 Mha are considered as compacted, 11 Mha as waterlogged and 5 Mha as subsiding (Figure 7.2). In comparison with erosion and chemical degradation, physical soil degradation comprises only 4.2 % of the total human-induced soil degradation (Oldeman *et al.* 1991).

However, the total amount of land affected (83 Mha) equals the combined land area of Norway, Sweden and Denmark. In addition, as previously mentioned, not all aspects of physical degradation are accounted for. Surface sealing is not included, just as the physical degradation of soil due to overbuilding. The damage done by ground sealing, foundations, sewers, cable trenches etc. can be seen on derelict sites and abandoned urban land. Given the rapid expansion of cities with their metastasing suburbs and post-industrial land demand, the amount of land consumed by sealing and overbuilding is expected to increase further.

In addition, GLASOD only records subsidence of organic soils, i.e. peat and marshland. However, underground mining has led to widespread subsidence and devalued precious land resources. The British and German Coal Mining Districts, the Transvaal Goldfields of South Africa, and the Pennsylvanian Mining District of the United States are only a few examples of areas which have experienced regional ground degradation due to mining. In most cases it is almost impossible to predict the

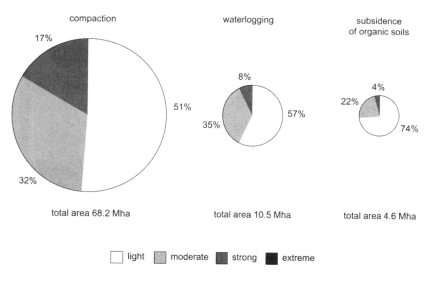

Figure 7.2 *Degrees of physical degradation: compaction, waterlogging and subsidence due to sealing and overbuilding (after Oldeman et al. 1991)*

subsidence potential and the time when subsidence will actually occur. Therefore, certain areas have become unfit for human use for an undefined period of time. As well as the extraction of minerals, the extraction of fluids causes subsidence. The region of Wilmington close to Los Angeles is subsiding due to oil extraction and the city of Bangkok is subsiding due to water extraction. Even the pumping of gas prompts subsidence, for example around the City of Groningen in the Netherlands.

7.3 Compaction

Compaction is a process of structural change of the soil. It is caused by the application of loads at the surface of the ground. Soil compaction is accompanied by an increase in soil bulk density, a reduction of pore space and a reduction of permeability. According to the German Environmental Protection Agency (Lebert *et al.* 2004) damage of the soil structure in the sense of soil compaction occurs when

- soil air capacity (the pore space only occupied after heavy rainfall or noncapillary porosity) decreases to $< 5\%$
- water conductivity becomes $< 1.16 \times 10^{-6}\,\mathrm{m\,s^{-1}}\ (\equiv 10\,\mathrm{cm\,day^{-1}})$
- bulk density is classified as 'dense' according to field tests.

Compaction may be a result of natural processes. For instance, during the ice ages the ground was compacted by massive glaciers. Layers of sediments that were subsequently eroded have also compacted the ground. Soil compaction can still be traced by means of oedometer tests, which produce stress-strain curves indicating the intensity of preloading (Chapter 4). Soils may also compact during earthquakes; alternatively, they may liquefy as examples from Norway, Sweden, Alaska, Canada and other countries demonstrate. In the latter case, sensitive clays (quick clays) have liquefied just as loose, evenly grained and water saturated sands (quick sands) do when loaded dynamically.

Natural ground compaction is often associated with the deeper soil profile while human-induced soil compaction is primarily concerned with the upper soil profile.

- Ploughing produces a compacted zone beneath the plough base that is referred to as plough sole.
- Overgrazing and trampling of cattle and other heavy livestock significantly compacts the upper soil horizons.
- On forested terrain, lumbering compacts the ground when felling trees, working the trunks and transporting them to stocking areas and out of the forest.
- Activities of leisure such as skiing, cross-country riding and mountain-biking on preferred paths compact soil especially in areas where soil profiles are shallow and the environment is fragile.

The effect of external loads on soil was studied by Valentin Joseph Boussinesq (1842–1929), who expressed the propagation of radial stresses σ_r as a function of the force F acting on the surface, the distance d, the angle α from the vertical and the concentration factor v

$$\sigma_r = \frac{vF}{2\pi d^2}(\cos \alpha)^{v-2}$$

as depicted in Figure 7.3.

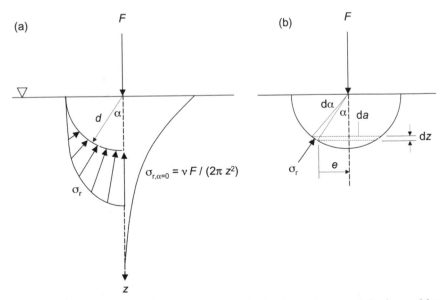

Figure 7.3 (a) Propagation of stresses below a point load (z = depth) and (b) the equilibrium of forces to derive Boussinesq's stress propagation equation

131

$v=3$ refers to an isotropic medium in which the modulus of elasticity is constant with depths; $v = 4$ refers to an anisotropic medium where the modulus of elasticity increases with depth. With increasing concentration factor, the stresses converge along the line of action of the external force (Figure 7.4a). This is the case for cohesive soils, whereas coarse soils have a smaller concentration factor. We may therefore conclude that compaction increases with:

- increasing surface load and
- increasing clay content

We also know from the Proctor Test (Chapter 4) that compaction increases

- with repeated loading

and that maximal compaction is reached at

- optimal soil water content.

Ground compaction is thus governed by loading, clay content and water content. It is also known that as the width of the load application increases, the zone of influence reaches deeper into the ground (Figure 7.4b).

Compaction due to repeated ploughing at the same depth has been investigated for many years (for example, Unger and Kasper 1994). Field studies in Poland indicate that the bulk density below the plough zone may increase by up to 20 % or more (Stojek 2004). According to this study and investigations carried out in Lower Saxony (Lebert and Schäfer 2005), soils prone to compaction are loamy silts and sands, gleysols and clay-dominated soils. Similar results follow from a study conducted in Zimbabwe (Tsimba *et al.* 1999), where sandy and loamy soils are identified as susceptible to compaction.

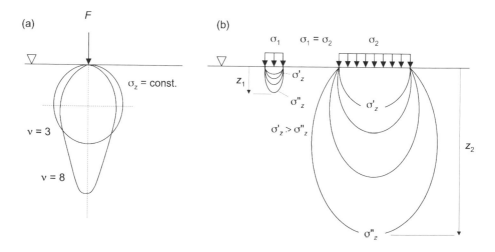

Figure 7.4 (*a*) *Contours of equal stresses for* $v = 3$ *and* $v = 8$ *and* (*b*) *with increasing width of load application, the zone of influence reaches deeper into the ground*

The introduction of heavy agricultural machinery has led to the compaction of even deeper soil zones. Depending on the wheel pressure and the soil moisture, compaction may affect the soil to depths of more than a metre (Figure 7.5). Wet soils compact about five times deeper than dry soils (BFS 2002).

Pastureland may become heavily compacted due to animal trampling. Although this type of soil compaction does not reach as deep as plough compaction, it may severely degrade the soil, especially on slopes. In the Northern Tundra of Russia, vast deer pastures have been identified as compacted (Stolbovoi and Fischer 1997). Similar observations are reported from other parts of the world, for example from the Mediterranean, where soil covers are thin and already eroded to a considerable degree.

Forestland may suffer soil compaction if heavy machines are brought in for felling trees and preparing them for transport. Temporary stocking of lumber may lead to soil compaction, especially if lumber is repeatedly stocked at the same site. The wheel pressure of lumber trucks also causes compaction.

Besides farmland, pasture and forest land, natural land may also suffer considerable compaction. The thin and delicate mountain soils are compacted by hikers in the summer and skiers in the winter. Grading machines used to smooth rock and soil on ski-slopes cause considerable damage. Cross-country riding and mountain-biking has degraded natural land, especially around cities and holiday resorts. On land exposed to outdoor activities features of compaction and erosion complement and enhance each other. Compacted patches and courses often serve as nuclei for successive soil degradation, difficult to control and to remediate.

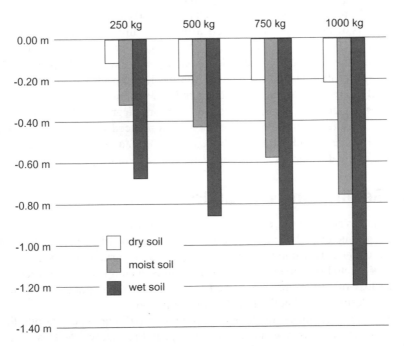

Figure 7.5 *Soil compaction as a function of soil moisture and wheel pressure for agricultural machinery (after BFS 2002)*

Soil also compacts under buildings, roads and other man-made constructions. However, since the upper soil horizons are usually removed before a construction is erected, the compaction mainly affects the C-horizon i.e. the parent material. Compaction due to structural loading is expressed in engineering terms as settlement. In order to avoid damage to the structure, settlement, especially differential settlement, has to be kept low. Therefore, soil may even be deliberately compacted to improve the building ground, either superficially with vibratory plates and rollers (grid-rollers for coarse soil and sheepsfoot rollers for fine soils) or to greater depth, for example by dropping weights from crawler cranes or by sinking vibrators into prepared holes that are successively filled with coarse aggregates (Bell 1993b). This type of ground compaction will remain even after the building is dismantled. Ground conditions on abandoned and cleared land are thus fairly inhomogeneous and certain measures have to be taken when land is recycled for redevelopment (Chapter 9).

Soil loses its functions because of compaction, especially if the solum is affected. First of all, the amount of pores and fissures within the compacted zone is reduced. This imposes on the capability of the soil to hold water and thus reduces the field capacity. The infiltration of rainwater may be reduced by more than half (Goudie 2000), prompting an increase in runoff and thus an increase in erosion. Due to compaction, less water and less oxygen are available in the rhizosphere. Roots may not be able to penetrate the compacted zone to reach water and nutrients. In addition, microbial activities slow down and the metabolism of plants is corrupted. As a result, the productivity of plants deteriorates and consequently the vegetation cover weakens, giving rise to serious erosion problems.

The oxygen and carbon dioxide exchange capacity between the soil and the atmosphere is disturbed, which adversely affects the climate. Groundwater resources are not sufficiently recharged and excess water accumulates above the compacted zone that may become waterlogged. Consequently, more water will run off and wear down the soil profile. Due to the increase in runoff, compacted soils also contribute significantly to the development of sudden and catastrophic floods.

It is estimated that about 68 Mha of soil are degraded due to compaction worldwide. This refers to about 3.5 % of all human-induced soil degradation. Of these 68 Mha, 51 % are considered as lightly degraded, 32 % as moderately degraded and 17 % as strongly degraded (Oldeman et al. 1991); see Figure 7.2.

7.4 Sealing and overbuilding

Sealing and overbuilding are the processes of covering ground permanently with engineering structures. The major cause of sealing and overbuilding is urbanisation. Since the upper active horizons of the soil profile are usually excavated before a building is erected or a road paved, urbanisation is accompanied by partial or even complete loss of the solum (the A- and B-horizon). Urban sprawl is thus directly linked to a destruction of both farmed and natural land.

In post-industrial times urban land consumption still continues, despite the fact that the population density is decreasing, as depicted in Figure 7.6. This is due to a global decrease of members per household and an increase in life expectancy i.e. an aging society (Chapter 3).

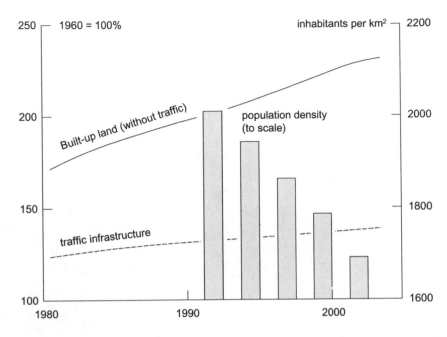

Figure 7.6 *In Germany, land consumption has increased while population density has fallen (after BBR 2005)*

In Europe, up to 10 % of the total land surface is sealed or overbuilt (EEA 2003). Extreme sealing rates are recorded from many cities where urbanisation has been consuming the periphery since industrialisation began. On abandoned industrial terrain, depicted for example in Figure 7.7, sealing rates remain high while ecologically intact land is claimed by new industries. Furthermore, due to an intensification of tourism, the overbuilding rate along the Mediterranean coast has increased dramatically. It is expected that this trend will continue even though the population remains stable in Europe (EEA 2003). In addition, sealing and overbuilding rates increase in the fast-growing economies of Asia and Eastern Europe.

Sealing and overbuilding creates a special soil type generally referred to as urbic anthrosols (Chapter 2), which is characterised by

- surfaces paved with asphalt and concrete slabs
- foundations and foundation fragments made of concrete, reinforced concrete or masonry
- pipes, sewers and tanks
- power and telephone cables
- allochtone (brought in) refill soil
- refuse and waste material
- soil and water contamination

(Genske 2003).

Figure 7.7 *Former Lauchhammer coking plant, Germany (photograph courtesy of Ariane Ruff)*

Soil loses its functions because of sealing and overbuilding. The loss of the solum brings about the loss of all ecological functions. Plants and animals are deprived of their natural living environment. The infiltration is disrupted and groundwater is no longer recharged. During storm events, the runoff drastically increases on built-up land, resulting in overcharging of the rivers and prompting sudden and disastrous flooding. In the UK, about 2 million homes are located in regions prone to flooding (Sunday Times 2000). Their economic value is of the order of 350 billion euros. It is expected that within the next 20 years, another 3.8 million homes will be added to this category. Every year, the floods in England and Wales alone cause an economic damage of around 1 billion euros (EEA 2000). The autumn flood of 2000 certainly outscored these statistics. In Germany, the Danube-Flood of 2002 affected 340 000 people, of which 100 000 had to leave their homes. The damage done by this single flood event is estimated to be in the order of 23 billion euros (Losch and Cordsen 2004a).

When land is sealed and overbuilt, almost all biological activities are brought to a halt. The resulting deterioration in biodiversity in urban environments is obvious. In addition, sealing and overbuilding obstruct the oxygen and carbon exchange capacity between the soil and the atmosphere. Since the upper active soil horizons are destroyed, carbon can no longer be sequestered, giving rise to a further degradation of the climate. In addition, the microclimate is altered due to the fact that sealed and overbuilt land heats up much more than natural terrain.

7.5 Subsidence

Subsidence is a gradual or sudden downward movement of the Earth's surface. In contrast to settlement that is caused by external forces such as structural loads, subsidence is caused by the extraction of resources, notably

- minerals
- fluids and
- gases.

Subsidence may cause a variety of problems including structural damage to buildings and infrastructure. When mining began in the German Ruhr District in the 19th century, wastewater could no longer drain from the subsiding settlements, which led to severe health problems and even epidemics. Subsidence may also disturb natural watercourses and aquifers. It may also provoke waterlogging and the development of lakes. Close to the shoreline, subsidence increases the risk of flooding.

As well as subsidence, the extraction of minerals, fluids and gases may also trigger earthquakes. For example, in 1967, tapping the Koyna aquifer in India induced an earthquake of magnitude 6.5 on the Richter scale and in 1976, tapping the Gazli Gas field in Uzbekistan produced an earthquake of magnitude 7.3.

7.5.1 Extraction of minerals

In Chapter 3 we stated that there are two ways to extract minerals: surface mining (quarrying) and subsurface (deep) mining. For surface mining the overburden is completely removed. Unfortunately, the upper active horizons are often not properly stored, which complicates the remediation work once mining has ceased. In some cases the overburden is even dumped and thus destroyed biologically. Surface mining brings about subsidence only indirectly when groundwater is pumped to keep the pit dry. A shallow, in most cases regular, subsidence trough develops, an aspect that will be discussed later.

Subsurface mining, on the other hand, is always accompanied by subsidence. At the surface, spontaneous development of lakes, drowning of canals and cracks in buildings may be observed (Figure 7.8). Subsidence may be

- regular i.e. associated with a gentle deformation of the surface or
- irregular i.e. associated with sharp surface breaks and caved-in crown holes.

The type of subsidence depends on the geology, the depth of the working and the mining technology. Shallow mining typically causes irregular subsidence, whereas subsidence produced by deep mining is usually more regular. Nevertheless, even deep workings may generate irregular deformation of the surface if the geology and the mining technique prove to be unfavourable.

Depending on the material to be mined, a variety of mining technologies have been developed. In longwall mining (Figure 7.9), panels of mineral (usually coal) are extracted along a working face between two parallel roadways. With proceeding longwall, the worked-out space or *goaf* is left behind, which subsequently caves in.

(a) (b)

(c) (d)

Figure 7.8 *Features of subsidence: (a) a flooded forest, (b) a waterlogged crop field, (c) a flooded open-air sewer, German Ruhr District and (d) a damaged church, Czech Republic (photographs (a), (b) and (c) courtesy of Peter Drecker and photograph (d) courtesy of Robin D. Koster and Siefko Slob)*

This mining technique typically causes regular subsidence that can be predicted with empirical diagrams, with the width and the depths of the working being input parameters as well as the angle of draw i.e. the angle that defines the limits of the subsidence trough, dependent upon the local geological conditions.

In room-and-pillar mining (Figure 7.10), the mineral is extracted with linear workings that cross each other, leaving rock pillars to support the roof of the mine.

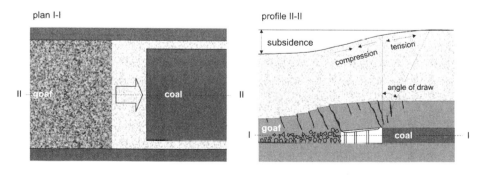

Figure 7.9 *Longwall mining*

The failure of the roof and the following subsidence consequently depend upon the depth of the workings, the geometry and thickness of the pillars and the geological conditions. If a pillar is overstressed it fails and the neighbouring pillars have to carry the additional roof load. These may in turn fail in a domino fashion, thus causing a sudden subsidence of the surface. When a mine is decommissioned, the pillars are usually reduced in diameter to maximise the exploitation before closure. Pillar robbing, however, significantly reduces the stability of the mine, prompting sudden and unexpected subsidence as the example of the Middleburg Mine in South Africa demonstrates, where about 215 ha are successively subsiding. Another mechanism of failure is the caving-in of the roof of the workings, especially at the crossings of the galleries. Broken rock bulks up in the working until the caving process ceases. In shallow workings this may lead to the sudden development of a crown hole. However, even in deeper mines the cavity may work its way up to the surface, especially if the caved-in material is eroded. In South Africa, vast regions like the Witbank Coal Field have been degraded due to room-and-pillar mining and have become unsuitable for human settlement (Bell et al. 2000).

Many other regions in the word are also affected, including the mining districts of Asia, the Americas, Africa and Europe. All these mines pose subsidence problems, not only while the mine is exploited but also after decommissioning, when the natural groundwater level re-establishes. Once a mine is flooded, stabilising and destabilising forces rearrange within the mine and processes of rock alteration and erosion commence.

Additional mining techniques have been developed e.g. solution mining to extract soluble minerals such as rock salt. For this technique, water is pumped into the soluble formation and brine is extracted. Solution mining creates irregular voids that may close slowly over many years, or rapidly cave in. Irregular subsidence and sinkholes may develop as well as regular subsidence, as observed in the brine fields of Windsor in Canada, where brine is extracted at depths of around 300 m causing subsidence of up to 8 m. Similar examples can be found in the Netherlands and other countries with exploitable soluble minerals. The historic brine fields in Cheshire, England, provide an especially well-documented example of the effects of solution mining on the surface (Bell et al. 2000). The possible subsidence volume is assessed by calculating the volume of brine extracted.

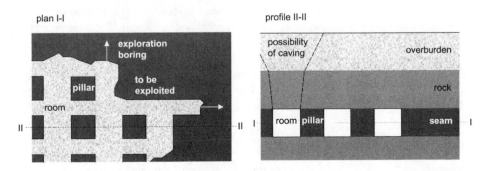

Figure 7.10 *Room-and-pillar mining*

Some of the subsidence phenomena are associated only indirectly with the working itself. For instance, the dewatering of the gold mines in the Far West of South Africa has lowered the groundwater table by more than 300 m. The karstic limestone formation was drained, causing the fillings of the cavities to shrink. The voids that developed begin to cave in without any warning. In August 1964, a collapsing void created a depression 30 m deep and 55 m across, swallowing a home and all its inhabitants (Goudie 2000).

Derelict shafts, previously used to lower the miners, raise the mineral and ventilate the mine, constitute considerable subsidence risks. Once a shaft has been decommissioned, it is in most cases only provisionally filled, sometimes even left open and simply covered with a concrete top. As long as the location is documented, the shaft remains recognisable as a danger zone. If the documentation is lost, an enormous risk arises for any redevelopment work. In the German Ruhr District, for instance, much material on mining commissions was lost during World War II. Although it is estimated that between 10–20 000 shafts exist in this densely populated region, only about half of them have been located. Shafts fail without warning every year.

In karst regions, underground voids develop naturally without any human intervention. Karst phenomena are encountered in all soluble rock formations including limestone, gypsum and rocksalt. In 1905, karst cavities were discovered when the Hales Bar-Dam at the Tennessee River (USA) was constructed. Both the construction time and costs quadrupled but, in spite of the effort, the subground of the dam proved permeable due to karst cavities. The dam finally had to be abandoned and a new dam was built 11 km up river. Karst cavities also led to major problems when the Santa Cruz Campus of the University of California was extended, the highway of Maluenda (Spain) was paved and when the tunnels for the high speed ICE-train in the Bundesland Hesse (Germany) were constructed (Genske 2005).

Subsidence at the Middleburg Mine, South Africa (Bell *et al.* 2000)

Mining operations at the Middleburg mine in the Witbank Coalfield, Mpumalanga Province, South Africa, began in 1908 by the room and pillar method. The coal seams in the Witbank Coalfield were formed in an epicontinental environment and occur within the Vryheid Formation. The Vryheid Formation forms the mid-part of the Ecca Group, which, in turn, is part of the Karoo Supergroup. This formation consists of sediments deposited in shallow marine and fluvio-deltaic environments in which coal developed from peat, which accumulated in swamps and marshes. The formation consists primarily of sandstones, siltstones, mudstones and shales. As the northern margin of the coalfield is approached, the sediments thin and the Vryheid Formation rests unconformably on the Transvaal Supergroup, the Waterberg Group and volcanic rocks associated with the Bushveld Igneous Complex. Minor dolerite dykes and sills are common. These have burnt and devolatised the coal seams in certain areas.

Room and pillar mining began in 1908. The average pillar size left behind was 6 × 6 m and the average room width was 7 m, so the extraction ratio was approximately 60 %. The average mining depth was in the region of 18–20 m, with a mining height of 2.5 m. From 1908 to 1947, when operations at the mine ceased, a total area of some 1700 ha was undermined using the room and pillar mining method.

Pillar robbing began in the late 1930s and continued until 1946. The original 6 × 6 m pillars were quartered, leaving four smaller pillars at each corner. The remnant pillars comprised approximately 25 % of the original pillar. This meant that the extraction ratio increased from 60 % to about 90 %. Any smaller pillars (4 × 4 m) were robbed on two sides leaving approximately 33 % of the original pillar intact. Pillar robbing occurred over an area of approximately 215 hectares.

There is little evidence that suggests that the original mining method caused subsidence in the area. This is understandable since mines with extraction ratios of up to 70 % are often relatively stable and, as remarked, the extraction ratio was around 60 %. Subsidence originated in the late 1930s (see Figure 7.11) when pillar robbing was practised, the extraction ratio being increased substantially. Pillar robbing increased the load that the pillars had to carry by as much as one third. Robbing also changed the shape of pillars by reducing their width significantly while the pillar height remained constant. This technique resulted in a reduction in pillar strength to the extent that many pillars could no longer support the overburden stress. Failure occurred when the ratio of pillar strength to vertical stress became less than 1.

The failure of a single pillar increases the stress on surrounding pillars, causing them to fail in domino fashion. The surface subsidence caused by multiple pillar failures is generally several hundred square meters in extent and the collapsed areas are often bounded by near vertical sides. Surface tension cracks around the outer edges of the collapsed areas are typically 20–80 cm in width and can extend in length for up to 100 m. In addition, some burning pillars collapsed and weakened the integrity of the remaining pillars, leading to further multiple pillar failure and related subsidence.

Figure 7.11 *Collapse of a pillared area in workings at Middleburg Colliery, Witbank Coalfield, South Africa (photograph courtesy of Fred Bell)*

141

7.5.2 Extraction of fluids

The withdrawal of fluids changes the state of equilibrium in the ground. Groundwater pumping lowers the groundwater table with the consequence that the drained soil loses its buoyancy. The soil begins to compact under its own weight.

Natural springs and watercourses were not satisfying the need for water even in ancient times, and people began to dig wells. With increasing populations and especially with the mechanisation of agriculture, the demand for water increased. Industrialisation and the relentless growth of cities instigated an even higher demand for water. As a result, many aquifers all over the world are overpumped; overpumping is of course accompanied by subsidence (see Figure 7.12, for example). For Tokyo, Mexico City and California, total subsidence rates of up to 10 m have been registered. Similar subsidence problems are reported in Europe, for instance from the Po-Basin in Italy or the region of Flanders in Belgium. In Bangkok, the MGL (Mitigation of Groundwater crisis and Land subsidence) project was launched, aimed at cutting down illegal water abstraction. This was apparently in vain as an ever-increasing number of wells have been drilled, especially in the suburbs where people are not willing to wait for permissions. Given the fact that the City of Bangkok is located in the Menam-Basin, only a few metres above the see level, exhausting the aquifer has become a delicate issue.

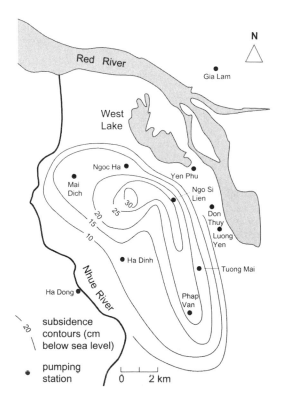

Figure 7.12 *Subsidence due to water extraction in Hanoi, Vietnam (after Nguyen and Heim 1995)*

142

Many other cities overpump their aquifers including Jakarta, Hanoi, Kanagawa and Beijing. In the San Joaquin and the Santa Clara valleys of California, water was pumped from the start of the 20th century, triggering subsidence of up 10 m (see Figure 7.13). Hundreds of millions of dollars were spent on revamping flood control structures and constructing a new one, as well as repairing structural damage at buildings and infrastructure and restoring wells. It was not until the introduction of the North-to-South Water Delivery System in the mid-1979s that subsidence was halted. Within this sophisticated water supply system, the Sacramento-San Joaquin River Delta plays a key role. There, however, we encounter another form of subsidence, this time prompted by the drainage of organic soils.

Organic soils (histosols) form from dead plants that accumulate under peat conditions i.e. saturated, anaerobic (oxygen-poor), acidic conditions under which decay is delayed. Peat can accumulate at rates of almost $2\,mm\,yr^{-1}$ until accumulation balances destruction. Profiles of up to 15 m may develop under certain conditions (Barrow 1994).

More than half of the Earth's peatland is located in Russia and Canada, almost half of which is frozen for about 40 weeks of the year. In moderate and warm climates,

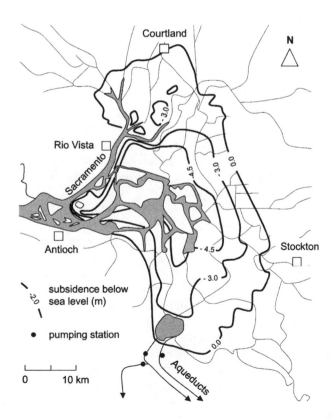

Figure 7.13 *Subsidence due to the drainage of marshland of the Sacramento-San Joaquin River Delta. The depth below sea level of the former peatland, now poldered, is indicated (after USGS 2000)*

peatland is often drained to make the land arable and suitable for human settlement. When peat soils are drained, they become exposed to the air and microbial decomposition accelerates. Organic matter is readily transformed into carbon dioxide and water, causing a reduction in volume, thus prompting subsidence. Furthermore, the drained soil begins to consolidate due to its own weight, an effect that also increases the subsidence rate.

The Sacramento-San Joaquin Delta is just one example of many regions in the world where subsidence of organic soils can be observed. The Dutch lowlands and the English Fenlands, where the Holme Fen Post was sunk in 1852 to register the subsidence, are other examples. At that time, the top of the cast-iron column was level with the peat surface. Today, it rises about 4 m above ground level. It is estimated that about 5 Mha of organic soil are subsiding worldwide due to human intervention i.e. about 0.25 % of all human-induced soil degradation. Of these 5 Mha, 74 % are considered to be lightly degraded, 22 % to be moderately degraded and 4 % to be strongly degraded (ISRIC 2005).

As well as the withdrawal of water, oil pumping causes considerable subsidence. The Wilmington Oil Field close to Los Angeles produces subsidence rates of more than 20 cm yr^{-1}. From 1928 to 1971, the surface subsided about 10 m (Goudie 2000). The Belridge concession, also in California, subsides up to 40 cm yr^{-1} (van der Kooij 1997, Fielding *et al.* 1998). Subsidence has been recorded from many other parts of the world e.g. the Romashkino Oil Field in Russia and from the oil platforms of the North Sea.

Subsidence of the San Joaquin Delta, California, USA (USGS 2000)

From the late 1800s, the peatland at the delta of the Sacramento River and the San Joaquin River was drained to create about 3000 km^2 of highly productive farmland. Since the conversion of peatland into farmland triggered subsidence of 2–8 cm yr^{-1}, almost 1800 km of levees had to be built. Today, a great part of the former peatland is below sea level, in some places up to 8 m below (see Figure 7.13). Complex drainage systems keep the farmland dry.

The delta, however, is not only used as agricultural land. It also receives about 40 % of the runoff of California and 50 % of California's total stream flow. It provides drinking water for two-thirds of California's population, and five times as much again is utilised for agriculture. The leveed farmland, especially the western seaward section, is vital to restrict the inland migration of brackish groundwater. A saltwater intrusion into the western polders would reduce the quantity of available freshwater, forcing the authorities to pump groundwater from the San Joaquin and the Santa Clara aquifers, which are already over-pumped. This would prompt further subsidence in an already subsiding area. Since the early 1890s, dams have failed more than 100 times. A sequence of dam failures would tip the water-exchange balance and more saltwater would intrude the delta. Unfortunately, dam failures are becoming more likely, since the ongoing subsidence makes the levee system increasingly vulnerable to floods and earthquakes. The State of California therefore plans to flood some of the polders with freshwater to restore the former wetlands whereas other sections will be filled with mineral soil and dredged material to slow down peat oxidation and raise the land surface.

7.5.3 Extraction of gas

Land also subsides when gas is extracted. For instance, when the Gazli Gas Field northwest of Bukhara (Uzbekistan) was tapped in the 1960s, subsidence rates of up to 2 cm yr^{-1} were recorded despite the fact that water was injected into the reservoir to counterbalance the loss of volume (Adushkin *et al.* 2000). Another example is the Mackenzie Delta in Canada, for which predicted subsidence due to gas extraction is depicted in Figure 7.14.

Since the 1950s, gas has been extracted in the Dutch Province of Friesland, prompting subsidence of a few decimetres. However, as well as the subsidence due to gas extraction, additional mechanisms that increase the subsidence rate have to be taken into consideration. Since the organic soils of the region were drained many decades ago, they are also subsiding due to self-compaction and peat decay. It is expected that the peatland will subside another half-metre within the next 40 years (Schokking 1995). In addition, a concession to carry out solution mining has been granted recently. Large volumes of salt (Zechstein) will be extracted from a depth of about 3000 m, which will accelerate the expected subsidence rates. Given that sea level is expected to rise due to global warming (http://www.ipcc.ch), the Dutch

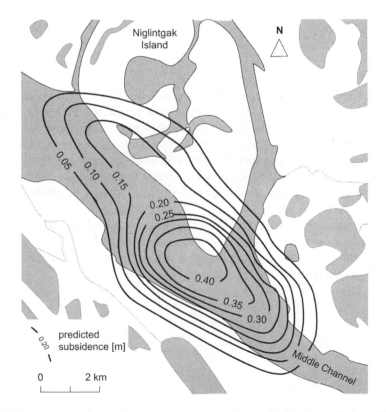

Figure 7.14 *Anticipated subsidence due to gas extraction in the Mackenzie Delta of Canada (after Tait and Hunter 2004)*

engineers are now facing a demanding challenge since the Province of Friesland is already located at, and at places below, sea level.

7.6 Waterlogging

Waterlogging occurs when more water enters the soil than it can drain. The excess water accumulates until it floods the surface. In many cases, only the root zone is affected and no direct indication of waterlogging can be detected at the surface.

A waterlogged soil typically shows aspects of gleysols, that is, soils with permanent or temporary wetness near the surface. The process of gleying is characterised by the mobilisation of reduced Fe- and Mg-compounds under anaerobic conditions that are either removed laterally with moving groundwater, or rise with capillary water above the water table where they segregate as mottles and concretions. A distinct mottling within the first 50 cm of the profile is thus indicative for gleying.

Waterlogging may occur as a natural process, for example:

- where a shallow clay layer impedes infiltration after heavy rainfall or
- in permafrost regions where the soil is frozen at a certain depth and is consequently impermeable.

On the other hand, waterlogging may be a consequence of human activities, for example due to:

- soil compaction or
- subsidence prompted by the extraction of minerals, fluids and gases.

A waterlogged rhizosphere impedes the plants from taking up oxygen. The soil structure declines and fungal diseases may start spreading. This sets in train a modification of the vegetation, as certain types of plants may expire while water-tolerant plants invade the affected terrain. On farmland the crop yield usually declines, bringing about considerable losses in revenue. In warm climates, water patches may serve as breeding ground for insects, promoting the proliferation of malaria and similar diseases. Buildings and structures cannot be safely erected on soils prone to waterlogging.

It is estimated that about 11 Mha of soil are degraded due to waterlogging worldwide, or about 0.5 % of all human-induced soil degradation. Of these 11 Mha, 57 % are lightly degraded, 35 % are considered moderately degraded and 8 % are strongly degraded (ISRIC 2005); see Figure 7.2.

III Remediating land

Approaches and techniques to investigate ground conditions are introduced. Secondly, remediation measures and techniques for land degraded by erosion and by chemical and physical impacts are presented. Finally, initiatives of land protection are outlined and a general strategy of sustainable land management is introduced.

8 Investigation

After presenting the general strategy of site investigation, the distinction between direct, indirect and fuzzy information is discussed. Approaches to associate and combine this information are introduced and methods of visualising the results are presented.

8.1 Strategy

The goal of site investigation is to obtain as much information as possible within a reasonable budget and period of time. The method of obtaining this goal has three basic steps:

- desk study
- field reconnaissance and
- field investigation.

The information collected during these steps is assembled and harmonised on thematic maps, based on the same reference map, depicting the *status quo* of the site. Usually, the *status quo* map is a topographic map providing morphology, watercourses, a simplified vegetation pattern, official land use and all existing buildings, roads and other man-made structures. Also given are geological outcrops, mining sites and quarries. The *status quo* map thus provides the basic information on the site that is complemented with thematic information.

An important aspect of collecting data is the facility to simplify and idealise information. The lucidity of a site investigation campaign suffers when too much information is included on the map, most of which is of minor importance and thus causing confusion. It is therefore important to filter out information not relevant to the map topic, and to reduce the remaining information to the essential aspects so that the presentation is clear and succinct.

Documentation and monitoring work complement the site investigation campaign. For example, the state of a site must be documented before a project has begun, in order to record any impact on the environment. Certain aspects like the concentration of contaminants in the groundwater, recovery of vegetation on degraded terrain or subsidence due to groundwater extraction may have to be monitored for a long time after a site investigation is complete.

8.1.1 Desk study

During the first step of a site investigation, the desk study, all information available from records is collected. The evaluation may include (Dodt *et al.* 1993):

- Textual records such as reports, publications and regulations that might be of importance for the site under investigation.
- Non-textual records i.e. spatial documents such as topographic maps, historical maps, geological maps, geomorphological maps, hydrogeological maps, supplementary plans (soil maps, land use maps, ecological maps, etc.), aerial photography and related remote sensing imagery.
- Oral information i.e. evidence of eyewitnesses.

Textual records are available at public offices, libraries and archives. Private companies formerly charged with investigation work may have supplementary information as well as local historical societies, environmental groups and initiatives. Textual reports and publications describe the state of the site under investigation and explain certain peculiarities e.g. given soil conditions. Of special importance are statistics commenting on mass flows, such as the amount of fertilisers applied per hectare or the amount of noxious substances produced and handled at a site currently considered as contaminated.

Although non-textual records (see above) are usually not detailed enough, they nevertheless provide a general overview of the geological development, the existing landforms and the groundwater conditions. Geological guidebooks and comments complement the maps. Another valuable source of non-textual information are thematic maps such as soil maps, land use maps, mining maps, ecological maps, etc. On industrially used, built-over land, historical maps provide information on how the land was used previously, where former buildings were located before they were dismantled and where the ground is possibly contaminated. Aerial photographs and remote sensing imagery constitute another valuable source of information. They indicate geological structure, vegetation, land use, soil types, erosion and drainage patterns, for example.

Verbal reports (possibly communicated as written reports) may complete the desk study and explain certain phenomena observed. They include descriptions of historical land use practices, former installations, accidents that have led to ground pollution, untypical soil types, hidden disposal sites, etc. Oral information usually reflects a subjective view that has to be evaluated as relative and fuzzy. Although still useful, it should be processed as such with the tools of fuzzy logic, explained later.

Historical evaluation of the coal mine Minister Stein
(Genske 2003, Genske et al. 1992)

The coalmine Minister Stein used to be one of the most productive deep coalmines in the German Ruhr District. Mining commenced in the latter half of the 19th century and a variety of processing facilities and chemical plants were founded in the direct vicinity. However, when it became cheaper to buy coal on the world market in the 1980s than to mine it at depths of around 1000 m, the mine went out of business. The site was abandoned and became typical industrial wasteland, too contaminated for potential investors. However, the still intact infrastructure of the immediate neighbourhood and the proximity to major Autobahns made it attractive again.

In the late 1980s, European funds were made available to remediate the site. Eight million euros were drawn from the European Fund for Regional Development to support the project and plans were made to integrate Minister Stein into the *Internationale Bauausstellung IBA Emscherpark*, an international building fair. The site became a prominent example of the conversion of derelict terrain into high quality development land.

In order to assess the hazards related to its former use, the first step was to analyse the production processes that took place on the site. In addition, a *status quo* worksheet illustrating the present situation on the site was prepared (Figure 8.1). Historic maps and aerial photographs were scanned (Figure 8.2) and significant features were imported into the *status quo* document. Based on the evaluation of the manufacturing processes, historic maps and building permits, the historic development of the site could be reconstructed i.e. installations and facilities suspected as potential sources of contamination could be identified. Digital layers were prepared representing historical stages of development. Finally, a synthetic map with the relevant historic information was produced (Figure 8.3).

8.1.2 Field reconnaissance

Field reconnaissance is aimed at gaining further information on the terrain and its degree of degradation. All relevant information gathered during the desk study is confirmed and new findings are recorded. A field reconnaissance campaign is essential to avoid misinterpretations and to optimise the state of information. It involves three basic steps:

- sampling
- testing
- and mapping.

Portable hand operated sampling devices have been developed, including hand augers and bailers (sunk into the ground to obtain single, well-disturbed samples), flight augers (rotated into the ground to obtain continuous, well-disturbed samples) and tube-type samplers (pushed, hammered or vibrated into the ground to fill a tube with a continuous, disturbed sample). The mechanically driven split sampler is a rapid sampling device that collects fine-grained soil along a notched probing rod. All these devices yield disturbed samples from which the type and the components of the soil as well as the depth of transition to other soil strata can be inferred.

In order to determine soil properties like unit weight, porosity, consistency, plasticity, organic content, water content, water retention capacity, strength and deformation characteristics, soil samples of a high quality have to be extracted. Undisturbed samples can be retrieved by gently pushing a special soil sampler into the ground. The sampling device is then carefully excavated and sealed on-site to avoid any disturbance and change in water content.

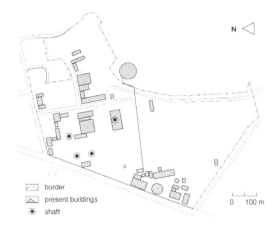

Figure 8.1 *Status quo worksheet of the Minister Stein site*

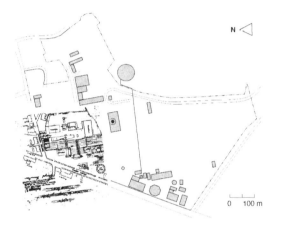

Figure 8.2 *Scanned historical building permits of Minister Stein site*

Figure 8.3 *Minister Stein historical development reconstruction*

152

Trial trenches allow a direct insight into the ground conditions. They can be excavated readily and at a low cost. The soil profile with its different strata can be mapped and undisturbed samples can be taken. If the bedrock is close to the surface, rock types, petrography and stratigraphy as well as the degree of fracturing can be determined. The depth of an unsupported trial trench is, however, limited to about two-thirds of the field assistant's height (i.e. about 1.20 metres on average), to enable safe exit in case of the trench caving in. Two people have to be present when mapping a trench, one remaining outside the trench. Trenches must not be excavated along the foot of a slope since this may trigger a slope failure. Instead, trenches may be dug perpendicular to the strike of the slope, with one trench at the foot, one at the crest and one in the middle, permitting a good insight into the weathering and erosion pattern. On industrially degraded land, noxious chemicals may be encountered in a trench, calling for health precautions and special protective gear.

The location of the sample and its depth has to be recorded, since every sampling point constitutes exclusive point information that will, as explained below, be associated with other point information to compose a thematic map. As well as this, the date and exact time of sampling, the equipment used, the company in charge and the person responsible all have to be noted on the sampling report. All circumstances that may have disturbed the quality of the sample, such as adverse weather conditions, have to be recorded.

As well as sampling, measurements and tests may be carried out to determine physical and chemical characteristics. Two types can be distinguished:

- field tests that are carried out at the spot e.g. penetration tests, percolation test, runoff measurements, etc.
- laboratory tests that are carried out on samples collected during the field campaign e.g. the measurement of the water retention capacity, the cation exchange capacity, the shear strength, the contamination, etc. (Chapter 4).

All the information gained through sampling and testing is visualised on maps. This information must be sorted and filtered to produce maps that are clear and to the point. A *status quo* map depicting the topography of the site serves as a reference map for all thematic maps produced. A thorough field reconnaissance campaign includes mapping

- geology: the types of sediments and rock and their state of weathering and fracturing
- geomorphology: the types of landforms and their relation to the given geology
- soil: the types of soil and their relation to the geology and the landforms
- hydrogeology: the hydraulic properties of the ground including the depth to the groundwater table and the permeability of the ground
- vegetation: the types of plants present and their relation to the given ground conditions
- human impact: the degradation of the ground due to the present and former land use, including aspects of erosion, chemical degradation and physical degradation.

In many projects, mapping is pursued incorrectly and the consequences are mistakes in development and remediation work. It is therefore recommended to include mapping campaigns in project work in spite of the fact that additional time and money is spent. The benefit gained from the additional information achieved usually outscores the additional work.

The field reconnaissance report concludes the second step of site investigation. This report explains the findings of the first two steps i.e. the desk study and the field reconnaissance. It includes, as well as a definition of the goal of the investigation, a thorough description of the geological and hydrogeological boundary conditions, the types of soil present and their degree of degradation. The tests carried out are commented upon and their results are summarised; detailed results are listed in appendices. Reasons leading to ground degradation are explained and illustrated with thematic maps that are all associated with the *status quo* worksheet. The field reconnaissance report concludes with recommendations for possible soil amelioration measures and suggests further investigation work to be carried out in the next step: the field investigation.

8.1.3 Field investigation

Field investigation aims at optimising the level of information on the terrain and its degree of degradation. It involves the same three steps as distinguished in field reconnaissance:

- sampling
- testing
- and mapping.

The spectrum of possible field investigation methods is vast and every year new investigation techniques are invented. Field investigation work is much more focussed than the desk study and the field reconnaissance campaign. Since questions on the state of the site have mostly been answered, only investigation work complementing the findings of the first two steps is necessary. In general, this phase of site investigation involves working with specialised equipment to carry out more advanced tests.

The preceding investigation work has identified locations where the ground has to be explored in more detail. Again, trial trenches are excavated to explore the ground conditions. Exploration borings may be sunk to retrieve samples from deeper ground. Lysimeters may be installed to monitor the hydraulic behaviour of the upper soil horizons. Geophysical surveys may be carried out to examine certain geological aspects. All these investigations and many more allow the description of the ground conditions in great detail.

The field investigation concludes with a report explaining the goals of the field campaign and the findings. At this stage, it may also become necessary to indicate additional measures of monitoring. For instance, the concentration of a certain contaminant in the groundwater may have to be monitored in order to determine the potential of natural attenuation. The equipment necessary to carry out the monitoring work is included in the report, as well as the recommended sequence and time span of sampling.

8.2 Information

In the following, the most important investigation techniques are grouped with regard to the type of information they yield. A distinction is made between

- direct information
- indirect information and
- fuzzy information.

During both desk study and fieldwork, all three types of information are collected, sorted, evaluated and visualised. In the following, the means to gain this information is presented in more detail.

8.2.1 Direct information

Direct information directly addresses a target value relevant to the site investigation campaign. Direct information can be obtained from

- maps and remote sensing
- field observations and measurements as well as
- laboratory tests.

The interpretation of maps, aerial photography and satellite scans is typically carried out during the desk study phase. The information gained is confirmed in the field with direct observation and measurements. Samples are collected for laboratory analysis.

Direct information is directly related to the point where the observation is made, the test carried out or the sample taken. In comparison with indirect and fuzzy information, direct information is specific and unambiguous. It is therefore the most important class of information. However, the process of gaining direct information may be both time and cost intensive. Therefore, indirect and fuzzy information is just as important to complement the information gained with direct observations and measurements.

8.2.1.1 *General ground conditions*

Maps and remote sensing

Direct information on the ground conditions can be acquired directly from maps including:

- Topographic maps, depicting rivers and lakes, land use and man-made structures, the morphology of the surface, etc.
- Geological maps and profiles, depicting the distribution, type and age of sediments and rocks, their deformation and fracturing due to tectonic forces and other aspects to facilitate the understanding of the geological boundary conditions.
- Hydrogeological maps, depicting aquifers and water protection zones, the depths to the groundwater table and other information to understand the hydrogeological boundary conditions.

- Soil maps, depicting the type and the distribution of soils, their utilisation and other information to understand the pedological boundary conditions.
- Other specialised maps including mining maps, geotechnical maps, ecosystem maps, etc. focussing on information not included in the basic maps mentioned above.

These maps are provided at public archives and services as well as other organisations and private companies. Maps are complemented with:

- Aerial photos and satellite scans provide information on general geological structures such as sediment basins and fault systems. The direct interpretation of the relief allows certain geological aspects to be recognised. For example, Figure 8.4 depicts an aerial photograph of the Murnau region in Bavaria, from which direct information on land use and geological features can be extracted. Stereoscopic viewing can provide additional information on geological features. In order to view aerial photos in three dimensions, two overlapping photographs are examined by means of a stereoscope. The stereopair of Oconto County in Wisconsin (Figure 8.5) allows the esker that stretches east of the river in the northsouth direction to be studied, indicating that the land was once glaciated.

As well as direct information, a large amount of indirect information can also be extracted from maps and remote sensing imagery, as explained below.

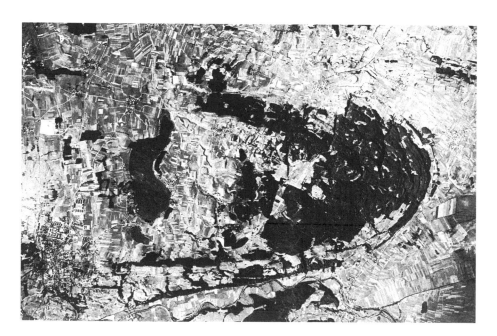

Figure 8.4 *Aerial photo of the Murnau region in Bavaria. Direct information on land use, geological features such as folding and faulting and glacial elements can be extracted (source unknown)*

156

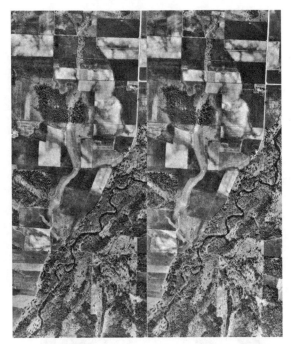

Figure 8.5 *A stereopair of Oconto County in Wisconsin. Stereographic viewing allows the clear recognition of the esker that stretches east of the river in the north-south direction, indicating that the land was once glaciated*

Direct inspection

Maps, aerial photos and satellite images are not usually detailed enough to allow a thorough site characterisation. It is therefore essential to confirm the extracted information in the field. The general ground conditions can be observed at denuded outcrops, along natural cuts and in trial trenches. Pedological information on the soil types, the soil horizons and their depth below the surface can be collected as well as geological information on the type of soil and rock, the degree of weathering and fracturing and other diagnostic parameters. In addition, samples can be taken from soil, groundwater and soil air.

Directly exploring deeper soil layers and the bedrock can be achieved by sinking core borings. Rotating core barrels are advanced into both soil and rock by the cutting action of the circular cutting shoe (Figure 8.6). In order to protect the inner core barrel from the drilling fluid and cutting action, double-tube samplers and even triple-tube samplers are employed. Wire-line techniques are applied to lift the inner barrel while the outer shell stays in the ground to be recharged with an empty barrel. Figure 8.7 depicts careful storage of cores in preparation for further inspection. As information may be lost during storage and transportation, a precise core log (Figure 8.8) is recorded onsite, documenting: 1 general information, 2 depth, 3 measurements (permeability, strength properties, RQD, etc.), 4 samples (undisturbed/disturbed), 5 geology, 6 description and 7 signatures and remarks (see Figure 8.8). Since drilling is

Figure 8.6 *Cutting shoes for core sampling*

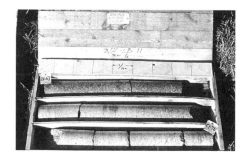

Figure 8.7 *Rock cores*

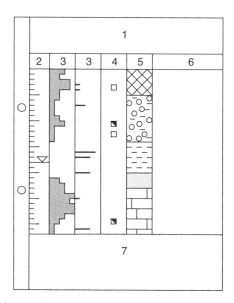

Figure 8.8 *General layout of a core log*

expensive, it is usually only carried out at locations that have been identified as suitable during the desk study and the field reconnaissance campaign.

Drill holes may subsequently be utilised as wells to collect hydrogeological information on the depth of the groundwater table and its change over time, the direction of groundwater flow, its velocity and the permeability of sedimentary strata below the active soil horizons. In addition, the water quality can be examined and any contamination detected. Drill holes may be complemented with piezometers (Figure 8.9), small-diameter wells to derive maps of the groundwater table. A free groundwater table or phreatic surface marks the separation of the saturated from the unsaturated zone (Figure 8.10). Piezometers and wells allow the observation of the depth to the groundwater table as well as the construction of ground profiles with water bearing strata or aquifers and strata of low permeability or aquitards. Aquitards may cover aquifers in such a way that a free groundwater table cannot develop. In this case, the piezometric surface defines the groundwater table. An artesian groundwater table is observed when the piezometric surface rises above the ground surface (Figure 8.10).

Well readings and piezometric readings are utilised to draw smooth equipotential lines, also referred to as groundwater contours. The contour lines indicate in which direction the groundwater slopes and thus give the direction of groundwater flow. The flow direction can also be determined by means of tracer tests, allowing the velocity of groundwater flow to be determined. For this test, a tracer (dyes, salts or radioactive tracers) is injected to observe its migration with monitoring wells, where the concentration of the tracer is continuously measured. The median of the concentration curve i.e. the moment when 50 % of the tracer has passed the observation well, marks the injection event and thus permits the back calculation of the flow velocity (see Figure 8.11). This velocity is also called the absolute velocity, since it is smaller than

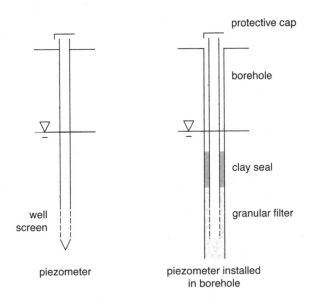

Figure 8.9 *Piezometer and well (simplified)*

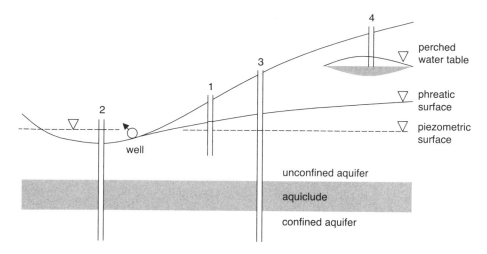

Figure 8.10 *Unconfined, confined and perched water tables with piezometer readings: piezometer 1 indicates the free water table or phreatic surface of the unconfined aquifer; piezometers 2 and 3 penetrate a confined aquifer and indicate its piezometric surface; piezometer 4 marks a perched water table*

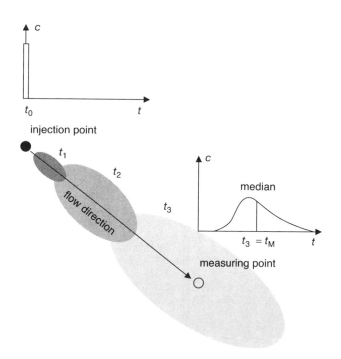

Figure 8.11 *A tracer test with an instantaneous tracer input at $t = t_0$ and a breakthrough curve at $t = t_3$; c refers to tracer concentration*

160

the real velocity within the pore channels and cracks and also differs from Darcy's velocity as measured in the laboratory. Lately, single borehole tests have been introduced to measure the flow direction and velocity by means of the drift of fines within the borehole. However, these tests only provide information on the flow situation in the direct vicinity of the borehole, which may in turn influence the outcome of the measurement.

The permeability of the soil may be determined from samples in certified laboratories. However, during a field investigation campaign, pumping tests and infiltration tests can be carried out by means of the already existing wells. They cover a much larger volume and are thus much more representative although more expensive (Chapter 4).

8.2.1.2 Erosion

The simplest method of directly measuring erosion in the field is to install measuring rods or erosion pins that are driven through a washer into the ground sufficiently deep so as to be uninfluenced by wetting and drying and the frost-thaw cycle (Morgan 2005, Schmidt 1983, Emmett 1965) (see Figure 8.12). The observation of the gap between the top of the rod and the surface of the ground is a direct measurement of soil erosion at this particular spot. If the washer becomes buried, the gap decreases, indicating that sedimentation is taking place. A recent refinement in the method is the introduction of a device referred to as a soil erosion bridge (Figure 8.12). A soil erosion bridge is a portable metal bar placed across two adjacent pegs which are firmly fixed in the ground (Hudson 1964). The vertical distance between the measuring bar and the ground indicates erosion and sedimentation processes. The measurements are regularly taken over a longer period of time, usually several years, to identify erosion and deposition trends.

Erosion troughs (Gerlach 1966, Seiler 1980) are designed to catch runoff and thus directly measure the quantity of fluvial erosion for a given plot. A distinction

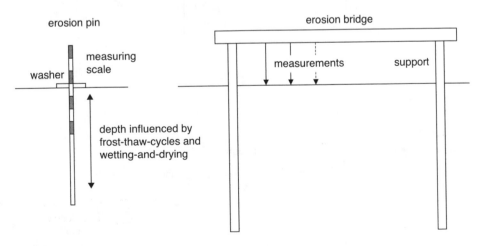

Figure 8.12 *Erosion pins and erosion bridge (simplified)*

has to be made between troughs that collect both water and sediments (e.g. Gerlach trough, Figure 8.13) and troughs that collect only sediments while the water is allowed to infiltrate into the soil behind the trough (e.g. Seiler trough, Figure 8.14). The latter type is usually preferred, since it is easy to install and at a low cost. The sediments caught in the trough comprise the eroded material coming from the plot ahead of the trough. In addition to the amount of soil loss, the composition of the eroded soil with regard to grain size and organic matter can be assessed and contaminants such as pesticides or industrial pollutants can be traced. Furthermore, the effect of any surface manipulation causing erosion, such as ploughing or the introduction of a hiking path, can be monitored in the same way as measures to control erosion.

A fully equipped research plot is usually bounded to avoid material transport from outside the defined area. As well as erosion troughs, additional measuring devices are installed e.g. a weather station with precipitation gauges, splash funnels and soil moisture detectors. In addition, lysimeters are installed to study the hydrological cycle in more detail and to measure aspects like infiltration, runoff, evapotranspiration and transport phenomena. A range of different types of lysimeters is depicted in Figure 8.15. A research plot allows a variety of different and refined measurements, however, its installation and maintenance is expensive and is therefore only feasible in the course of a research project.

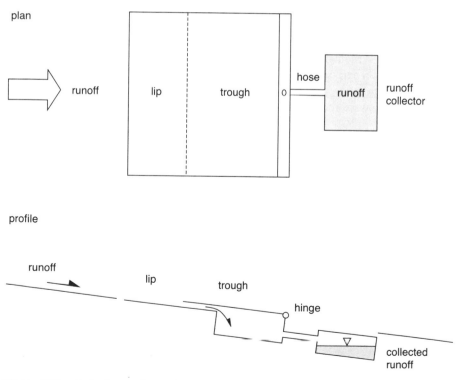

Figure 8.13 *Gerlach trough to collect the entire runoff (water and sediments)*

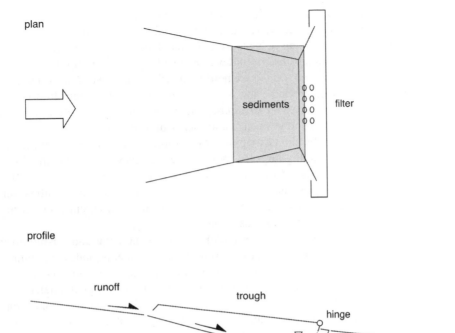

Figure 8.14 *Seiler trough to collect only the sediments (simplified)*

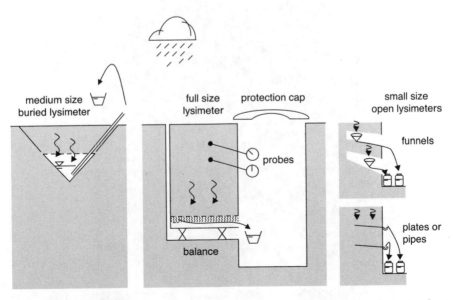

Figure 8.15 *Different types of tension-free lysimeters (simplified): medium size buried lysimeters; full size lysimeters; small size open lysimeters*

Dust and sand traps have been developed to directly quantify the effects of wind erosion. Although a proper aerodynamic design of such sediment traps has proven difficult because of the build-up of back pressure that deflects the wind from the trap, a number of sediment catchers has been developed including the Big Spring Number Eight Sampler and the Wilson and Cooke bottle sampler, which always faces the wind due to the attachment of a wind vane to the sampling mast (Morgan 2005). The sampling receptacles are fixed at different heights in order to sample a vertical sediment profile with regard to grain size and total sediment amount.

Aerial photos allow the direct mapping of erosion phenomena from gullies to complete drainage patterns. Figure 8.16, for example, depicts wind erosion in a loess deposit in Nebraska. A sequence of photos taken over a certain time permits a multi-temporal analysis of the development of erosion impacts and measures of mitigation. Aerial photos and satellite scans also provide indirect information, enabling spatial erosion to be modelled as will be discussed later.

Many attempts have been made to model erosion phenomena under laboratory conditions including rainfall simulation, runoff measurements and wind tunnel experiments. These experiments aim at determining which parameters govern erosion processes, for subsequent use in modelling real erosion phenomena. The advantage of such tests is that they are conducted under clear boundary conditions and therefore yield precise and reliable results. The disadvantage, however, is that the complexity of field conditions is difficult to simulate in the laboratory, which again calls in question the applicability and the robustness of the results.

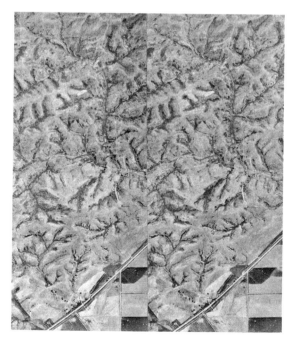

Figure 8.16 *Stereopair depicting gullies in a loess deposit in Nebraska*

8.2.1.3   Chemical degradation

Contamination

Direct information on the grade and intensity of contamination is provided by the sampling of soil, groundwater and soil gas. As in all sampling campaigns, tracing of contaminants has to be based on the preceding desk study. The type of samples taken depends upon the kind of pollutants expected. As mentioned in Chapter 6, the groups of contaminants include organic pollutants, inorganic pollutants and radionuclides. Organic pollutants can further be divided into non-volatile organics, semi-volatile organics and volatile organics. In addition, contaminants may be soluble or insoluble.

Soil samples may be taken from the surface, from a trial trench, from the bottom of a borehole or from a drill core (Figure 8.17), when tracing pollutants. When the sample is expected to be contaminated, special precautions have to be taken.

Groundwater samples may be taken from a well (Figure 8.17). Note that contaminants dissolved in the groundwater are mobile, just as for contaminated soil gas. Therefore, the interpretation of the information gained from the measurements only gives a temporary picture of the contaminant distribution. Usually, a sequence of samples is taken over a period of time, at the same sampling spot, in order to measure possible variations of the contamination.

Gas samples are taken when volatile compounds are suspected, for example above a VOC/SVOC-contamination (volatile/semivolatile organic contamination), close to a landfill or in the vicinity of a derelict mining shaft (Figure 8.18). It should be noted that the atmospheric pressure may considerably influence soil gas measurements.

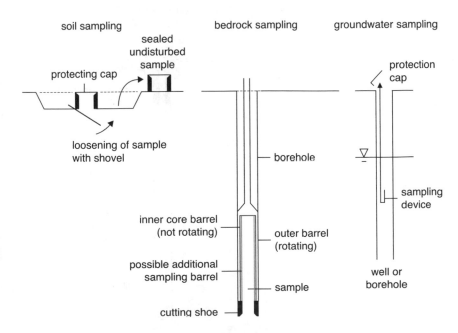

Figure 8.17 *Sampling of soil, bedrock and groundwater (simplified)*

Hand driven spikes yield weak and unreliable results and should therefore not be used. Today, gas wells can be installed easily and cheaply, for instance with window samplers or similar equipment. Samples have to be sealed immediately to avoid contact with air. The possibility of any chemical reaction between the contaminant and the sampling device as well as the storage receptacle should be avoided.

The samples are subsequently analysed to detect traces of possible contamination. Some measurements can be carried out directly in the field whereas more complex analyses have to be conducted in certified laboratories. Depending on the type of pollution, care has to be taken in order to avoid heath risks. The location and the time of sampling have to be recorded, as well as the sampling device utilised, the weather conditions and the person in charge.

Soil samples, groundwater samples and soil gas samples only provide point information for interpretation in order to derive maps depicting the spatial intensity of the contamination. Contouring routines are applied to interpolate the data obtained at the sampling points. However, as will be shown later, the kriging method provides results that are much more reliable.

A sampling campaign may either aim to detect a background contamination, a diffuse contamination caused by agrochemicals and wind blown contaminants, or a contamination source caused by a hidden dumping site or a dismantled industrial installation. Additionally, sampling campaigns are conducted to monitor remediation measures. The routines and details of sampling contaminated terrain are provided in national codes and regulations.

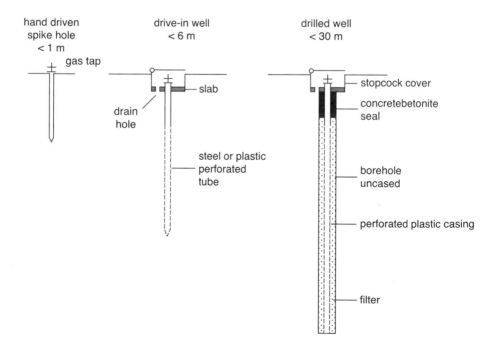

Figure 8.18 *Gas sampling (after Wilson and Haines 2005)*

166

Acidification and salinization

Direct information on the grade and intensity of acidification and salinization is provided by soil sampling. The methods to retrieve samples are the same as discussed above. The grade of acidification is determined by measuring the pH, whereas the grade of salinization is determined by measuring the electrical conductivity of the pore water (Chapter 4).

Solidification

Although the solidification of soils involves complex chemical processes that lead to hard pan development and lateritization, the detection of solidified soil layers is comparable with the detection of compacted soil layers. Direct means of measuring the grade of solidification is discussed below.

8.2.1.4 Physical degradation

Compaction

The method of quantifying compaction is measuring the resistance of the soil against the penetration of a rod. In the 1930s, Künzel licensed a testing device consisting of a simple steel rod that is hammered into the ground by means of standardised weights dropped from pre-defined heights. The number of blowcounts per penetration depth is a measure of the resistance and thus the strength of the ground. Penetration tests have been refined and standardised in many countries and are usually applied to inspect the bearing capacity of the ground when constructing buildings and engineering structures. Alternatively, cone penetration tests have been developed with a 60° cone driven into the ground at constant speed. The cone resistance and the friction along the outer sheath are measures of the strength of the soil.

Penetration tests are also applied to detect a possible compaction of the soil. If the resistance increases significantly at the same depth while the soil type remains the same, a compacted layer has been encountered. Simple compaction testers are just pushed into the ground with a meter indicating the pressure applied. More refined penetrometers have been developed with the cone being hammered into the ground and the number of hammer blows per penetration distance as an indicator of compaction. Since the hammer weight and the falling height remain constant, results can be compared and typified as in the case of the standard penetration test device used for engineering purposes (Herrick and Jones 2002), as depicted in Figure 8.19.

Sealing and overbuilding

Sealing and overbuilding can be recorded directly by conducting surveys, either terrestrial or by remote sensing. Terrestrial surveys are usually carried out for small areas like derelict industrial sites. The only sealing elements that are actually visible are those on the surface of the terrain, however. More may be buried in the ground such as bunkers or obsolete foundation plates. It is therefore essential to complement

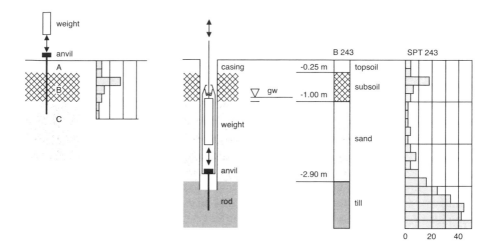

Figure 8.19 *Penetration test from the surface, in a borehole (simplified)*

such surveys with the information gained during the desk study. The history of the site is revealed with records and historic maps, verified by trial trenches and possibly borings. At this point, crisp information may have to be combined with less reliable i.e. fuzzy information. In addition, it is not always easy to decide whether a certain sector is actually sealed; it may only be sealed to a certain degree. Meinel and Hernig (2005) allocate a sealing grade of 100 % to impermeable areas and roofs, 70 % to semi-permeable terrain like paved sites, 50 % to green roofs and water absorbing areas like gravel beds and 0 % to meadows, gardens and natural land.

For larger areas, aerial photography and satellite images are utilised. Although aerial photos provide direct information about the sectors that are overbuilt, the sealing grade of surfaces like backyards or paved roads remains difficult to assess. Satellite scans add additional indirect information since they cover a much broader electromagnetic spectrum. Their utilisation will be discussed later in the context of indirect information.

On industrial wasteland the natural soil profile is usually disturbed. The depth up to which a disturbance can be expected varies according to the history of the site. Frequent features are:

- Foundations and foundation fragments made from concrete, reinforced concrete, or masonry. Usually flat footings are located below the depth that frost can reach during a severe winter. On the other hand, footings are normally located above the groundwater table. However, as well as flat footings other foundation elements such as wooden, concrete and steel piles may reach deep into the ground, well below the groundwater table.
- Pipes, sewers and tanks that are usually aligned with transport pathways and concentrated in the vicinity of former buildings.
- Power and phone cables that are used to connect dismantled facilities and installations.

- Allochthon granular soil that was built in to replace weak soil in order to improve foundation conditions.
- Waste material dumped in the vicinity of production units or at special dumping places.
- Soil and groundwater contamination.

The Wilhelmine Victoria case file (Berief and Krupop 1997, Genske 2003)

An example on how a terrain may be unknowingly degraded is the former coalmine Wilhelmine Victoria in the German Ruhr District. It was planned to convert the former industrial land by building a large number of private homes on it. Although the first homes were built without any problems, enormous foundations were suddenly discovered at a depth of only 80 centimetres, as shown in Figure 8.20. Nobody had expected such massive obstacles. Inquiries led to the conclusion that they belong to an ancient pithead building, which had been dismantled a long time ago. The foundations were left in the ground, since their extraction would have called for a considerable budget that was not available.

For fourteen days a team of workers tried to chisel away the foundations consisting of reinforced concrete—without success. Finally, the planners were forced to build the the home without a basement. Building permits had to be modified and the statics of the building had to be revised, causing delays and costs that so far have not been recovered.

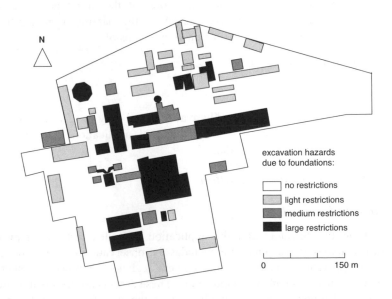

excavation hazards
due to foundations:

- no restrictions
- light restrictions
- medium restrictions
- large restrictions

0   150 m

Figure 8.20 *Map showing excavation hazards due to foundations (Berief and Krupop 1997)*

Subsidence

Direct information on subsidence can readily be obtained by surveying the surface. The survey is later repeated to monitor any changes over time. Traditionally, terrestrial surveys are carried out with survey points as reference marks. However, a large amount of data is required to obtain a statistically sound comparison. Acquiring such data is time consuming and costly. Today, satellite scans permit the measurement of the deformation of the surface in an inexpensive and rapid way. European Space Agency (ESA) satellites provide synthetic aperture radar (SAR) data. An interferogram can be calculated from a pair of SAR images, a function of both surface topography and displacement. A third SAR image, acquired at a later time, is used to eliminate topography and measure the surface deformation. SAR interferometry (InSAR) is used to provide high-resolution maps over large areas ideal for the remote measurement of land subsidence (Massonnet *et al.* 1997, Fielding *et al.* 1998, Buckley 2005 and others).

Waterlogging

Waterlogging can be directly detected as soon as water floods the surface. As long as the water level is below the surface, trenches and shallow wells may be employed to locate a possibly perched water table. In waterlogged terrain the vegetation will show signs of degradation—an indirect indicator as discussed below.

8.2.2 Indirect information

Direct information is complemented with indirect information. In contrast to direct information, indirect information does not represent the target parameter under consideration. Instead, effects that are caused by the parameter in question are recorded. Indirect information can be obtained by means of

- indicators and index tests
- maps and remote sensing
- geophysical sensing.

Indicators and index tests are readily available and easy to record. Geophysical and remote sensing has become a major field of application to detect ground conditions and the state of degradation.

8.2.2.1 General ground conditions

Indicators and index tests

The utilisation of indicators and the application of index tests are typical for pedological mapping campaigns. This includes the observation of the colour of the soil and its apparent structure, organic content, calcium carbonate content, soil moisture, root density and other soil indicators. These aspects are recorded in order to establish a soil diagnostics that eventually leads to the classification of the soil type and a pedological description of the soil profile.

Below the active soil horizons, grading curves derived from soil samples serve to determine the type of sediment. In addition, index values like the Atterberg limits, the consistency index and the activity index (Chapter 4) are utilised to assess the stress and strain characteristics of the soil. Penetration tests are carried out to correlate the number of blowcounts per distance of penetration with the friction angle of the soil and its compressive strength. During a cone penetration test, a cone is driven into the soil at constant speed (15–25 mm s^{-1}) in order to gain information on the type of soil and to determine whether or not the soil is cohesive. By means of a hand vane, which is pushed into the ground and turned to shear off the portion caught in the vanes, the shear strength of the soil can be assessed as well as its sensitivity (the loss of strength when loaded dynamically as, for example, in the case of an earthquake).

Another important source of indirect information are bioindicators. Certain plants prefer sandy ground, others fine grained and cohesive soils. Some plants grow on calcareous soil whereas others grow on salty or acid soil. *Pulmonaria obscura*, for example, marks loamy portions in sandy soils. Certain plants indicate the groundwater table. For instance, *Juncus effusus* indicate stagnant water whereas *Equisetum silvaticum* indicates shallow moving groundwater. Vegetation often marks changes in soil conditions and may even indicate faults in the bedrock that are water bearing and thus beneficial for the growths of plants.

Maps and remote sensing

As well as direct information, maps offer a variety of indirect information that the experienced interpreter can extract. The presence of certain features suggests further aspects relevant to the site investigation campaign. For instance, the pedologist would expect certain soil types along watercourses, in depressions and on hills as depicted on topographic maps. The ecologists would expect certain ecosystems based on the morphology of the terrain and the given climate. The engineering geologist would expect karstic bedrock in sectors where depressions cannot be explained with the given landforms or possible mining activities. Thematic maps and field investigations subsequently complement and confirm the conclusions that were drawn in an indirect way.

Remote sensing provides indirect information on the state of the ground for a wide range of applications. A distinction is made between passive and active sensors: the former record radiation reflected from the surface of the earth, while the latter send out electromagnetic waves at a particular frequency and record both the intensity of the returned signal and the time taken for the signal return (a function of sensor-earth distance).

Simple grey-tone photos are images of reflected visible light. However, they offer much more information. For example, the intensity of the grey tones infers the moisture content of the soil, from which the type of soil can be deduced. Dark grey tones denote moist and fine soils whereas light colours denote coarse, permeable soils. The texture of grey tones reveals the uniformity of the ground. Uniform grey tones denote a uniform geology whereas an irregular pattern suggests locally varying soil conditions. In addition, surface patterns visible on grey tone photos provide a wide range of indirect information as well as direct information on visible geological structures. The drainage pattern may accentuate geological features like faults hidden underneath a soil cover. The density of the drainage pattern gives an indication of the permeability of the ground and thus the possible types of soil present. Dendritic drainage patterns represent homogeneous ground conditions whereas pinnate drainage patterns indicate silt and loess soils. Since vegetation masks the natural grey tones the preferred period to take aerial photos is the time after snowmelt and before foliage. Coloured aerial photographs provide additional information on the geology and land use.

In addition to visible light, other regions of the electromagnetic (EM) spectrum may be utilised to discern geological structures (Lillesand and Kiefer 2003). It includes gamma rays, X-rays, ultraviolet radiation, infrared radiation, microwaves and radio waves, with the visible light (VIS) spanning about 0.4–0.7 μm. Gamma rays have wavelength $\lambda < 1 \times 10^{-5}$ μm and represent the most energetic form of EM radiation. Being a component of cosmic radiation, they are highly penetrating and difficult to focus. Since they are also emitted by naturally occurring radioactive elements they are used to map radioactive anomalies indicating geological structures. X-rays have wavelengths from 1×10^{-5}–0.01 μm. They are also difficult to focus and are thus rarely utilised to map geological structures.

Ultraviolet waves (UV) cover the region from 0.01–0.4 μm and are less energetic than X-rays. UV induces fluorescence in certain minerals like scheelite and calcium tungstate that can be mapped with UV-scans. Infrared (IR) radiation has λ: 0.7–300 μm and is employed to map active temperature contrasts (volcanoes, water temperatures, etc.). In addition, the way in which solar IR-radiation is adsorbed and reflected can be recorded in order to map material differences at the surface. Microwaves span from 0.0003–0.3 m and are utilised in Radiation Detection and Ranging (RADAR). Microwaves pass through materials with low dielectric constants (non-conductors) but are reflected by water-bearing soil and metal-bearing ore bodies. Both passively reflected solar microwaves and actively emitted microwaves are applied to map geological features. The latter technique is referred to as Synthetic Aperture RADAR (SAR) and serves to map surface topography as well as buried bedrock topography irrespective of cloud cover and vegetation. Radio waves cover the 0.3–300 000 m domain. They pass through non-conductors and are adsorbed by conductors. Since long wavelengths limit the size of objects that can be resolved, their application in remote sensing is limited (Spencer 1997).

Geophysical sensing

Geophysical sensing embraces a broad field of applications. Geophysical measurements aim at mapping property contrasts associated with target parameters, allowing their measurement, and include:

- gravity surveys
- magnetic surveys
- geoelectric surveys and
- seismic surveys.

Gravity surveys are conducted to map gravity contrasts. Since gravity is directly associated with density, gravity surveys track density anomalies. As early as the

1950s, gravity surveys were conducted along the Gulf Coast of the United States and Mexico to locate salt domes, which are often related to oil reservoirs. Salt has a much lower density than the surrounding material and is therefore easy to map. Newton's law of universal gravitation defines the gravitational forces F acting between two masses m_1 and m_2 separated by a distance r

$$F = \gamma \frac{m_1 m_2}{r^2}$$

where γ is the universal constant of gravitation ($6.673 \times 10^{-11} \ N\,m^2\,kg^{-2}$) as derived by Henry Cavendish in 1798. By introducing Newton's 2nd law of motion, that is, force equals mass times acceleration, the gravitational acceleration becomes

$$g = \frac{F}{m_2} = \gamma \frac{m_1}{r^2}$$

where m_1 is the attracting mass and m_2 is the attracted mass. Gravitational acceleration and gravitational potential

$$v = \gamma \frac{m_1}{r}$$

map gravity contrasts. The measurements are carried out with specialised gravity meters of the spring type and unstable gravimeters with oscillating amplification systems. The measurements have to be adjusted to elevation, topography, latitude and the mass residue to the datum reference ellipsoid (Bouguer correction). Gravity measurements are comparatively expensive and may yield convincing results (see Figure 8.21).

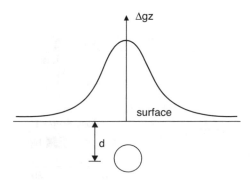

Figure 8.21 *Vertical components Δg_z of a density anomaly over a buried body; the shape of the curve yields information on the depth and geometry of the anomaly*

Detecting cavities on brownfields (Scheibe *et al.* 1997)

On a derelict industrial site in Berlin a gravity survey was carried out to detect cavities due to abandoned subsurface structures including cellars, tanks, and tunnels. As all surface buildings had been dismantled a long time ago and there was no sufficient historic information at hand, the survey was a prerequisite to redevelop the site. It revealed a number of large voids, possibly basements, underneath the former main plant. In order to redevelop the site it was necessary to dig up the voids and refill them with appropriate soil. The gravity survey, the results of which are depicted in Figure 8.22 (dark ⇒ low gravitation i.e. the presence of an underground void) helped to identify problem zones. An extensive drilling programme was therefore avoided and remediation work could progress in a cost-effective way.

Magnetic surveys are conducted to map magnetic contrasts. The method is one of the oldest techniques of geophysical prospecting. During a magnetic survey, the difference between the local magnetic field and the induced, remnant magnetisation of buried bodies is measured. The residue is mapped to visualise the shape of the

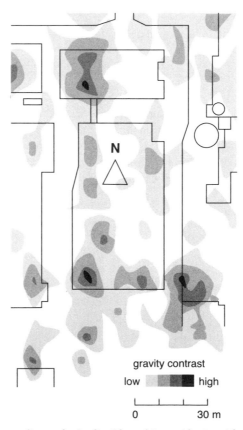

Figure 8.22 *Gravity anomalies at the Berlin-Oberschöneweide site with the present use of the site overlaid (Scheibe et al. 1997)*

anomaly. In analogy to Newton's law of universal gravitation, the magnetic forces F that acts between two poles p_1 and p_2 separated by the distance r is defined as

$$F = \frac{1}{\mu}\frac{p_1 p_2}{r^2}$$

where μ is the magnetic permeability. The magnetic field strength h is calculated as

$$h = \frac{F}{p_2} = \frac{1}{\mu}\frac{p_1}{r^2}$$

Magnetic field strength and magnetic potential

$$u = \frac{1}{\mu}\frac{p}{r}$$

are employed to map magnetic contrasts. The measurements were previously carried out with field balances. Later, more refined instruments were invented like the proton magnetometer which utilises nuclear magnetic resonance. During a magnetic survey, a base station records the intensity of the Earth's magnetic field while individual measurements are carried out along a raster of measuring points. The residues of the base station measurements and the individual measurements yield an image of magnetic anomalies that can be correlated with changes in the subground. Since only the residues are mapped, diurnal changes of the magnetic field do not influence the outcome. However, the survey may have to be interrupted if magnetic storms arise. Figure 8.23 represents the variation in vertical magnetic field strength Δh_z, a measure of the depth and geometry of any magnetic anomaly. Magnetic measurements are comparatively inexpensive and easy to carry out.

Geoelectric surveys are grouped into direct current and alternating current methods. In DC geoelectrics the resistance of the ground is recorded while an electric

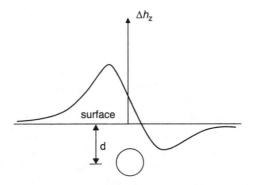

Figure 8.23 *Vertical components Δh_z of a magnetic anomaly over a buried body causing a magnetic contrast*

current is introduced by means of electrodes. Ohm's law relates the potential difference U to the resistance R and the current I applied

$$U = R I$$

With the introduction of the specific resistivity ρ Ohm's law becomes

$$\rho = \frac{R a}{l} = \frac{U}{I} \frac{a}{l}$$

where l is the length of the uniform conductor and a its cross-section. The resistivity of natural material varies considerably and depends mainly upon the pore fluids. A dry mineral matrix such as salt or granite has a very high resistivity whereas saturated loams and clays have a very low resistivity. An apparent resistivity ρ_a can be mapped by introducing a direct current with two electrodes and measuring the potential gradient

$$\rho_a(l,a) = \frac{U}{I} f_0; \qquad f_0 = \frac{\pi}{4} \frac{(l^2 - a^2)}{a}$$

The results depend on the layout of the measuring plot (Figure 8.24). Measurements of constant depth are conducted by shifting the measuring plot along a profile. Resistivity image profiling produces profiles along which geological and hydrogeological anomalies become visible. Vertical measurements are achieved by changing the distance between the electrodes. Vertical electrical sounding may be carried out with different configurations, notably the Wenner configuration with l being set equal to $3a$:

$$\rho_a(a) = \frac{U}{I} f_w; \qquad f_w = 2 \pi a$$

and the Schlumberger configuration with constant a and increasing l:

$$\rho_a(l,a) = \frac{U}{I} f_s; \qquad f_s = l^2 \frac{\pi}{4 a}$$

For each configuration model, curves have been developed which allow an assessment of the depth to geological boundaries such as the groundwater table or the depth to the bedrock (Figure 8.25).

In AC geoelectrics a signal is transmitted into the ground which causes induced currents to produce secondary magnetic fields. A receiver measures the residue of the two currents, which changes location, hence characterising the ground with respect to its electromagnetic resistivity. In electromagnetic imaging (EMI) the standard tool is a portable bar with a transmitter on one end and a receiver on the other. The distance between the transmitter and the receiver and the frequency of the signal governs the penetration depths. Ground Penetrating Radar is an electromagnetic reflection method

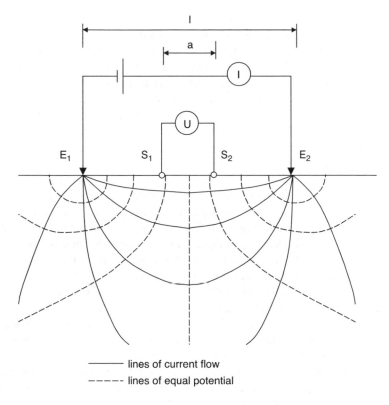

Figure 8.24 *Current electrodes (E₁, E₂) introduce a direct current that is measured with potential electrodes (S₁, S₂)*

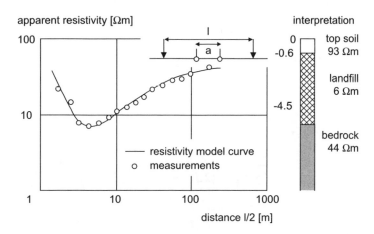

Figure 8.25 *Electrode configuration, resistivity curve and interpretation for vertical electrical sounding (after Vogelsang 1997)*

(EMR) of sending electromagnetic pulses via a radar antenna into the ground. Dielectric boundaries reflect the signals, which are recorded at the surface with a receiving antenna. AC geoelectrics is fast, inexpensive and extremely appropriate for locating geological features such as discontinuities and fault zones as well as anthropogenic impacts such as shallow cavities, derelict foundation fragments and other obstacles.

Other techniques which have been developed include induced polarisation (IP) surveys where the decay of an induced polarisation is utilised to map polarisation contrasts and self-potential (SP) surveys where naturally occurring electrical currents in the ground are detected.

Seismic surveys were introduced in the 19th century, when the first chair of seismology was established at the University of Göttingen, Germany. The observation of seismic waves produced by earthquakes permitted the description of the inner structure of the earth, the depth to its core and the thickness of its crust. Two types of waves can be distinguished: body (e.g. compressional or P- and shear or S-) and surface (e.g. Rayleigh and Love) waves. The velocities of these waves are governed by the elastic properties of the medium they travel through. Waves introduced artificially with hammer blows, explosives and vibrating machinery were first used commercially in the 1920s in North America and Germany. The waves travel into the ground from the shooting point to be reflected or refracted at geological boundaries. Geophones spread along profiles register the echoes of these waves. Since P-waves are the fastest, they are recorded first by the geophones. They are plotted on time-distance curves to reconstruct geological structures.

The ray path of waves reflected at geological boundaries can be analysed by applying Snell's 1st law (the angle of reflection equals the angle of impact) and the law of Pythagoras, as depicted in Figure 8.26. With

$$\overline{S'R}^2 = (2d)^2 + x^2$$

and

$$\overline{S'R} = v_1 t$$

where v_1 is the velocity in the first layer. It follows that

$$(v_1 t)^2 = (2d)^2 + x^2$$

that is

$$\frac{t^2}{\left(\dfrac{2d}{v_1}\right)^2} - \frac{x^2}{(2d)^2} = 1$$

and thus

$$t^2 = t_0^2 + \left(\frac{x}{v_1}\right)^2 ; \qquad t_0 = \frac{2d}{v_1}$$

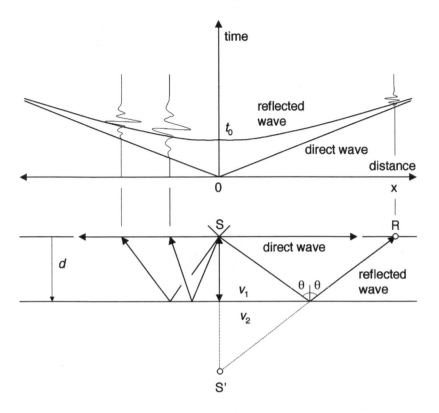

Figure 8.26 *Fastest ray path of reflecting P-waves and time-distance curve*

from which the depth to the reflector, the P-wave velocity in the first layer and elastic properties can be inferred. P-waves are, however, also refracted according to Snell's 2nd law:

$$\frac{\sin \theta_1}{v_1} = \frac{\sin \theta_2}{v_2}$$

as depicted in Figure 8.27.

At a certain distance from the shooting point, the refracted P-wave travels along the interface of the two layers while constantly producing new signals that are recorded at the surface. The first refracted wave touches the surface at the critical distance x_c from the shooting point. If the travel velocity of the wave in the second layer is higher, which is usually the case, the refracted wave arrives before the direct wave at the crossover distance x_{cross}. It thus follows that

$$t = \frac{x}{v_2} + \frac{2d}{v_1} \cos \theta_c$$

179

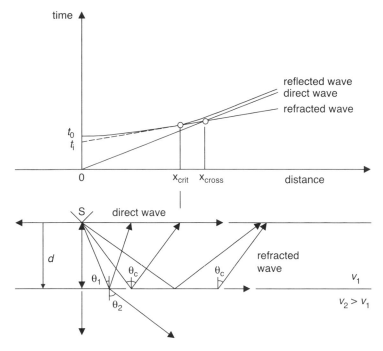

Figure 8.27 *Fastest ray path of refracting P-waves and time-distance curve*

yielding the depth to the refracting layer

$$d = \frac{1}{2} x_c \sqrt{\frac{(v_2 / v_1) - 1}{(v_2 / v_1) + 1}}$$

where v_1 is calculated from the arrival time of the direct wave

$$v_1 = \frac{x}{t}$$

and v_2 is calculated from the arrival time of the refracted wave

$$v_2 = \frac{x - x_{cross}}{t - t_{cross}}$$

Refraction seismics thus provide information on the depth of the refractor, the wave velocities in both layers and their elastic properties provided that the velocity in the deeper layer is higher. This technique is especially suitable for the investigation of shallow depths, as signals from reflected waves are often blurred by surface waves. It has also proven useful in the assessment of the depth to the bedrock. However, seismic surveys are usually expensive.

Geophysical surveys are also carried out in boreholes: self potential (SP) logging is used to locate fine soils with a low permeability; electrical resistivity (ER) logging to identify different strata; gamma ray (GR) logging is used to detect layers of high clay content; neutron (NG and NN) logging measures water content and porosity; gamma-gamma (GG) logging produces information on the density of the ground from which geological strata and discontinuities can be determined; acoustic logging yields information on the porosity and fracturing of the ground. In addition, shooting between boreholes reveals geological features between the shooting point and the geophones that are installed in the surrounding boreholes.

8.2.2.2 Erosion

Measuring the sediments leaving a catchment along a river over time is an indirect way of assessing the rate of fluvial erosion. The sediment yield of a river can be assessed using either special sediment samplers or by recording the turbidity of the water. Alternatively, the sedimentation rates in lakes and reservoirs can be evaluated to infer erosion rates. In this case, the trap efficiency of the reservoir i.e. the amount of sediments that escape via the spillway in times of floods has to be known. The assessment of sediment yields and sedimentation rates can be correlated to direct measurements made with field installations such as erosion rods and troughs set in place to estimate the soil loss per hectare.

Satellite imagery permits the study of sediment plumes in rivers and deltas caused by erosion. The dramatic effect of deforestation can be traced, especially after heavy rainfall. When the tropical cyclone Gafilo hit northern Madagascar on 7th and 8th of March 2004, astronauts at the International Space Station witnessed how eroded sediments migrated down the Betsiboka River towards the silted-up estuaries that were once frequented by ocean-going ships (UNEP 2005).

Remote sensing also helps to quantify the vegetation cover, which greatly influences susceptibility to erosion. For example, Thematic Mapper (TM) instruments have been launched on Landsat satellites to image the surface of the earth in 7 spectral bands (from 0.45–12.5 µm) at a spatial resolution of 30 m. Bands TM 3 (0.63–0.69 µm) and TM 4 (0.76–0.90 µm) are especially useful in the discrimination of vegetation as they record the reflectance values at the red and near-infrared bands of the EM spectrum respectively (Mathieu et al. 1997). To determine the density of green and living plants on a particular area of land, the normalized difference vegetation index (NDVI) is used.

$$NDVI = fn\left(\frac{TM\,4 - TM\,3}{TM\,4 + TM\,3}\right)$$

Since the chlorophyll in plants strongly absorbs the visible light (i.e. low reflectance in band 3) but strongly reflects the infra-red radiation (i.e. high reflectance in band 4), a high NDVI implies healthy vegetation and therefore low erodibilty. A low NDVI indicates a terrain most likely covered in dead or sparse vegetation, with an increased threat of erosion.

Thus, the cover factor C_f necessary for the RUSLE approach to modelling fluvial erosion (Chapter 5) can readily be assessed. In a similar way, the erosion control practice factor P_f can be assessed. In addition, the soil erodibility factor E_f can be inferred from the soil types detected as well as the S_f factor which reflects the morphology of the terrain. The erosivity factor I_f indicates the frequency and intensity of storms, determined from meteorological data. Processing remote sensing data with geo-information systems, incorporating the RUSLE approach, yields clear images of the erosion potential (Huth and Jürgens 1995, Jürgens and Fander 1993). In a similar way, parameters of the WEQ approach may be inferred from remote sensing imagery and integrated into a GIS-supported assessment of wind erosion potentials.

8.2.2.3   Chemical degradation

Contamination

There is a broad spectrum of indirect information which can indicate possible contamination. A straightforward approach is the analysis of mass-flows which have led to the contamination problem. For a diffuse contamination, the contaminant may have been introduced, for example, as a pesticide on farmland or as atmospheric fall-out of noxious exhausts. The contamination spreads over large areas in a diffuse way. A mass-flow analysis determines what quantity of contaminants have been released, and over what period of time. The contaminant input less the amount which has left the affected area, for example with the runoff, allows the amount of contaminants remaining on the site to be determined. Depending on the type of contaminant, the ground and climatic conditions, the contamination may decay with time, migrate to the groundwater or dilute in other ways, or simply remain immobile. Associating mass-flow analyses with transport models allows an assessment of diffuse contamination phenomena, although many input parameters may be estimates. It is therefore necessary to tune these models with field data.

Mass-flow analysis is also an essential tool to evaluate local contamination problems typical for derelict industrial land. The analysis starts with screening the production processes from the goods that were produced to the resources that were necessary to produce them. The analysis involves a thorough understanding of how all resources and products were handled, processed and stored and where all these processes took place (Hatheway 2002). Only after manufacturing processes and operational practices have been understood can the types of contaminants, their approximate amount and possible location become clear.

In this context, historical maps and information on the development of the site are of great value (Genske 2003). Historical maps and aerial photos are superimposed in order to reconstruct the development of the site and to locate possible spots of contamination. As explained previously, a multitemporal analysis is an integral part of the desk study and is routinely carried out when abandoned industrial terrain is investigated. In many cases, different generations of installations can be distinguished, interpreted with respect to their utilisation and their potential to cause contamination. Possible sources of contamination are sites of production, handling and deposition of noxious substances. A historical analysis is rarely complete e.g. certain editions of maps may have been lost during a war, and therefore details have to be confirmed from a thorough field investigation and complemented with direct, indirect and fuzzy information.

Additional sources of indirect on-site information are bioindicators. Specific plants tolerant to certain contaminants may invade the terrain, indicating the presence of a shallow contamination (Brooks 1998, Reeves and Baker 2000). They may be mapped to demarcate contaminated zones. For instance, more than 300 plants are listed as accumulating nickel. Some 50 of about 175 species of the genus *Alyssum* are nickel hyper-accumulators. *Viola* species accumulate zinc, *Agrostis* species and *Minuartia verna* accumulate lead and *Thlaspi* species accumulate both. Furthermore, *Thlaspi caerulescens* accumulates cadmium, whereas *Iberis intermedia* (Figure 8.28a) accumulates thalium. The grass species *Agrostis capillaries* and *Festuca ovina* as well as *Silene vulgaris* (Figure 8.28b) are typical plants tolerant of metals. Mapping plant communities is easily done and possible without a large budget; therefore a valuable tool in identifying sites of possible contamination.

Some sites are contaminated such that the natural vegetation suffers. Although many plants are *ubiquist* i.e. they tolerate a large range of soil conditions, high concentrations of contaminants may cause certain plants to disappear. In addition, the growth pattern may be corrupted. Changes in vegetation can be traced with satellite imagery, especially when the density of vegetation declines. In this context, the normalized difference vegetation index (NDVI) is of particular importance as it highlights the contrast between living plant tissue and areas where plants have died due to contamination.

In addition to remote sensing, geophysical sensing has also proven useful in tracking contaminants. Since certain contaminants change the conductivity and electromagnetic properties of the soil, they can be located with geoelectric surveys. Geoelectrics has also proven useful in mapping contamination plumes in the groundwater. Recently, much research has been carried out to optimise electromagnetic imaging and reflection methods (EMI/EMR). With magnetic surveys, buried waste dumps containing barrels and other metal containers can be located. Induced polarization (IP) methods have proven useful to spot deposits like

(a)   (b)

Figure 8.28 *(a) Iberis intermedia (photograph courtesy of Catherine Keller) and (b) Silene vulgaris tolerate metals*

galvanic sludge, ore processing refuse or municipal waste deposits. Gravity surveys have been applied to locate density anomalies on industrially degraded terrain that are caused by loose soil and voids next to massive foundations (Vogelsang 1997).

Acidification and salinization

Collecting indirect information on acidification and salinization is based on similar methods to those discussed above. The development of vegetation may already serve to map acidified or saline terrain. *Calcifuge* plants i.e. plants preferring acid soil, include for example *Erica, rhododendron* and some birch species. *Halophytes* like *Batis, Salicorna, Inula crithmoides, Crithmum maritimum, Salsola, Sarcocorinia* and *Crambe maritima* indicate salt-rich soils. In addition, the natural growth of plants may be disturbed when the pH becomes extreme and the salinity too high. This can also be traced by means of satellite scans which record such changes in vegetation. Since a change in pH and salinity evidently changes the properties of the pore water and thus the soil, geophysical methods may also be applied.

Solidification

Solidification of soil also has strong impacts on the development of plants. The penetration depth of the roots is limited and strong rainfall may lead to waterlogging that causes different kinds of root diseases. As mentioned before, barren land quickly develops when erosion reduces the soil profile to the solidified layer. This aspect can be traced with satellite scans as for regions affected by deforestation.

8.2.2.4 Physical degradation

Compaction

Compaction of soil is indicated by a number of index values including soil air capacity, water conductivity and bulk density. A physically compacted soil typically has soil air capacity < 5 %, water conductivity < 1.16×10^{-6} m s^{-1} (or 10 cm day^{-1}) and a bulk density classified as 'dense' according to field index tests. Since the field capacity and thus the infiltration rate are reduced, runoff quantities on compacted soils typically increase. Consequently, the amount of eroded soil also increases. Erosion measurements are thus an indirect way to monitor soil compaction.

Soil compaction can also be recognised by changes in growth patterns of plants. Since less water and oxygen is available within the rhizosphere, the health of the plant suffers. In addition, roots cannot penetrate the compacted horizon to obtain nutrients. As a result, the vegetation cover weakens and crop yield deteriorates, a phenomenon that can be observed during a field survey and with satellite scans. With respect to the latter approach which allows larger areas to be investigated, changes in the NDVI are of particular importance.

Within a compacted horizon the physical parameter of the soils are changed. Because of this, compacted layers can also be traced with geophysical tools, for instance electromagnetic imaging and reflection methods (EMI/EMR). These surveys are fairly specialised and are employed, for example, in the visualisation of ancient travel routes or places of historic, now derelict, buildings.

Sealing and overbuilding

Sealing and overbuilding is either mapped directly on-site or by means of aerial photos and satellite scans. The former method is usually employed when a detailed analysis is required, whereas the latter approach is applied for the investigation of larger areas. Since remotely sensed images can distinguish between vegetated and non-vegetated terrain, a rapid assessment of non-vegetated terrain is possible. However, a distinction between non-vegetated terrain and sealed terrain is not as straightforward, implying field data may be required in some cases to assist with or adjust the interpretation of the imagery.

With the installation of the European Environmental Agency (EEA) by the European Council in 1990, a European Environmental Information and Observation Network (EIONET) was set up. Under the CORINE Land Cover (CLC) programme, satellite images are analysed and interpreted to map the whole European territory into standard Land Cover Categories, of which there are 44. These categories range from wetlands to wooded lands, from agricultural land to artificial terrain. CORINE thus serves to recognize the type of land cover, identify changes over time and derive indicators that can further be processed to derive environmental risk maps (EEA 1999).

Sealing grades in the city of Berlin
(Senatsverwaltung für Stadtentwicklung 2006)

For west Berlin, a first assessment of the sealing grade was made in 1979 on the basis of aerial photographs. In the 1980s and 1990s the survey was integrated into an environmental information system based on blocks bounded by streets. In order to improve the precision of the assessment, satellite images were analysed. The sealing grades inferred from the Landsat TM scan deviated only a little from test sectors where the sealing grade was known precisely.

For east Berlin, the sealing grades were first assessed by means of older topographic maps and aerial photos. At densely built-over areas, aerial photos were difficult to interpret since the backyards of older multi-floor buildings were often obscured due to shadow. In these cases, the sealing grade was estimated on the basis of comparable buildings. Later, satellite scans were used to improve the precision of the evaluation. It was however difficult to distinguish between sealed areas and areas without vegetation, which made on-site confirmation necessary.

In domestic areas, the sealing grade ranges from 22–82 %; in the city centre it is estimated to be 77 %. Industrial areas are 74–89 % sealed, areas of infrastructure 63 %, municipal areas 26–83 %, traffic areas 70–96 % and parks, green areas and farmland 1–46 % sealed (see Figure 8.29). On average, the city of Berlin was found to be 34.7 % sealed.

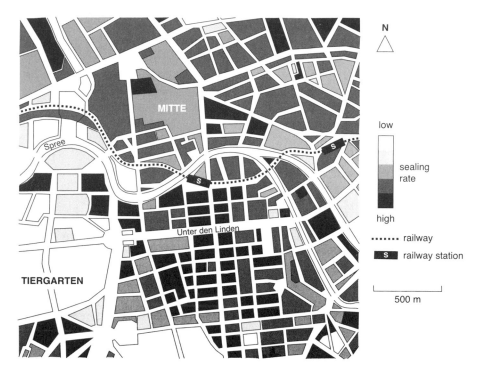

sealing rate

high

······· railway

S▮ railway station

500 m

Figure 8.29 *Sealing grade in central Berlin; darker grey tones indicate a higher sealing grade (after Senatsverwaltung für Stadtentwicklung Berlin 2006)*

Indirect information also provides information about former buildings and roads. Anomalies in growth patterns, remote sensing and geophysical sensing have enabled archaeologists to reconstruct the circular henge structure of Goseck, south of Berlin, for example. The monument believed to be a temple of the sun was discovered on aerial photos in the 1990s, which showed several concentric circular anomalies in vegetation of up to 70 m in diameter. Although timber was used to construct the palisade walls, which had rotted away long ago, the change in soil conditions caused changes in plant growth which mirrored the buried remains of the monument despite the fact that it was erected 6800 years ago (see Figure 8.30). The Goseck circles are thus much older than the megalithic circle of Stonehenge. Geophysical sensing revealed more details on the three entrance portals and the inner rings. A reconstruction of the solar constellation at the time the structure was built shows that two of the entries were aligned with sunset and sunrise at the time of winter solstice (Bertemes and Schlosser 2004).

Mapping bioindicators, remote sensing and geophysical sensing also produce valuable indirect information on derelict industrial terrain. Sectors that used to be overbuilt or sealed can be traced and obstacles buried in the ground such as foundation plates can be localised. In addition, caves, openings and tanks can be mapped as can cables, pipes and canals. Indirect information is especially useful on

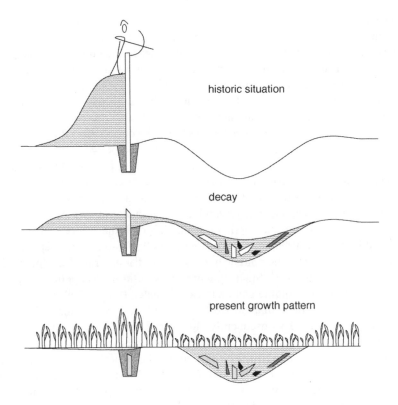

historic situation

decay

present growth pattern

Figure 8.30 *Historic structures leave traces in the ground that are indicated by changes in the growths pattern*

contaminated terrain, as excavation work may call for expensive safety measures and may possibly trigger the emission of the contaminant.

Subsidence

A similar approach can be applied to indirectly map areas of subsidence. Plants serve as bioindicators as soon as the water table approaches the surface. The rhizosphere might become permanently drawn, prompting a rapid deterioration of the health of the plants, which eventually die. This deterioration process can be observed on-site as well as with aerial photos and satellite data. At the fringe of a subsidence trough the surface slopes. Trees document the time when sloping began with a kink in the case of sudden sloping and bended, banana-like growth in the case of gradual sloping.

Cracks in buildings and man-made structures may indicate subsidence. Mapping the intensity of cracks provides information on how the subsidence trough develops. However, only structures that are similar with regard to size, time and type of construction should be compared to avoid misinterpretations.

Derelict shafts are especially dangerous since they may suddenly and unexpectedly cave in. They may not be recorded on topographic maps, especially if mining permissions were granted a long time ago and if some of the documents have been lost due to negligence, war or other events. Surface temperature is normally higher above shafts since they reach deep into the ground, and so shafts can be indirectly located by scanning the surface temperature. During winter, the snow on top of a derelict shaft may melt first. Airborne thermo-scanning reveals temperature differences at all times of the year. However, in densely populated regions too many other temperature sources blur the image and make interpretation difficult. If a shaft is suspected, it can nevertheless be located indirectly with a raster of soil gas measurements, provided that the gas emitted by the shaft differs from natural soil gas. This is the case, for instance, for coal mining shafts which emit methane, readily detected with soil gas measurements.

Calculating the mass balance for a mining district also allows an assessment of the possible subsidence. In the German Ruhr District, the overall volume of coal extracted from 1800–1990 has been estimated as 9.54 Gton, which represents 5 % of the world's production. The volume extracted is of the order of 7 km^3 plus 2.5 km^3 of waste rock i.e. about 9 km^3. Since stowing (back-filling) was applied only in special cases, the subsidence volume can be back calculated (Meyer 1993). From empirical measurements we know that 50–90 % of the thickness of the calculated seam manifests as subsidence. Consequently, the overall subsidence volume is estimated to be 6.0–7.2 km^3. The maximum subsidence recorded so far is about 25 m and compressive and extension tensile strains on the flanks of the subsidence trough have reached up to 10 mm m^{-1} (Szelag and Weber 1993). In certain regions of the Ruhr District, the groundwater table has touched the surface and wetlands and lakes have developed where forests and farmland used to be. In many cities, the groundwater table is kept artificially low to avoid flooding and the railroad tracks as well as the Autobahns and the drainage systems are monitored continuously for problems associated with subsidence.

Waterlogging

Waterlogging may be a consequence of soil compaction, subsidence and natural causes. It can therefore be mapped with bioindicators as described above since a drowned rhizosphere significantly weakens the vegetation cover and eventually kills the affected plants. Waterlogging can be readily observed via remote sensing. As well as this, indirect approaches to detect soil compaction and subsidence may also be relevant to demarcate areas prone to waterlogging.

8.2.3 Fuzzy information

In the case of a site investigation a large amount of information hinting at possible ground degradation has to be processed. However, only a small amount of this information can be regarded as precise, true or 'crisp'. Almost all statements incorporate some kind of linguistic softeners like "at the northern side there was *apparently* a *huge* landfill *some* twenty years ago". Figure 8.31 depicts both crisp and fuzzy descriptions of "a large tank of water". Although these statements are fuzzy,

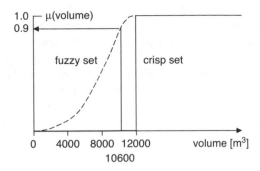

Figure 8.31 *Crisp and fuzzy description of 'large volume of a liquid waste tank'; a tank with a volume of 10 600 m³ belongs to the fuzzy set 'large' with a confidence value of 0.9*

they nevertheless include a certain degree of expert knowledge that might be useful to better understand the project site. But how can this information be quantified?

Lotfi A. Zadeh introduced in the 1960s an innovative concept to process information characterised by a limited truth-value. Although little understood in the beginning, the introduction of fuzzy logic has had a remarkable impact on the scientific communities and has become a valuable tool in the incorporation of incomplete information of a fuzzy nature.

Klir and Folger (1988) state that Aristotle already discussed the basic idea of fuzzy logic in his treatise *On interpretation*. Aristotle reasoned that future events must not be given a true or false status since they hadn't yet occurred. Their truth-value remains undetermined until they take place. He wrote:

"When [...] the reference is to universals, but the propositions are not universal, it is not always the case that one is true and the other false, for it is possible to state truly that man is white and that man is not white and that man is beautiful and that man is not beautiful; for if a man is deformed he is the reverse of beautiful, also if he is progressing towards beauty he is not yet beautiful."

The classical two-valued logic based on 'true' and 'false' thus has to be extended into a three-value logic that includes 'true', 'false' and 'indeterminate'. Even an n-valued logic can be developed which is, in fact, the basic idea of fuzzy logic. According to Klir and Folger (1988):

"Fuzzy logic is actually an extension of many-valued logics. Its ultimate goal is to provide foundations for approximate reasoning with imprecise propositions using fuzzy set theory as the principal tool."

Zadeh (1965) states that this concept

"...provides a natural way of dealing with problems in which the source of imprecision is the absence of sharply defined criteria rather than the presence of random variables."

Fuzzy logic thus provides a measure of vagueness and ambiguity. In contrast to a crisp set with a characteristic function $X_A(x)$ that determines whether x is a member of A, the membership function $\mu_A(x)$ indicates the degree to which x belongs to a fuzzy set A. Fuzzy sets are defined by their membership functions, which are therefore the core of the entire concept. Figure 8.32 depicts some commonly used types of membership functions. Standard membership functions approximate the ways humans linguistically interpret real values. Studies in psycholinguistics show that piecewise exponential functions perform better in complex systems (such as language and perception) than more simple linear systems (von Altrock 1995). Therefore, they are considered to describe environmental systems better, that is, more 'organically'.

Fuzzy sets can be combined on a logical IF-THEN base with AND, OR and NOT-operators, for example: IF x is small AND x is more or less true OR y is unlikely THEN z is expensive. Thus, fuzzy logic permits us to use fuzzy quantifiers (small), fuzzy truth-values (more or less true), fuzzy modifiers (unlikely) and fuzzy predicates (expensive). As can be seen from this discussion, fuzzy logic is a mathematically demanding tool. In this chapter, only the basic idea of the concept can be presented. More details are given in other textbooks such as Hanss (2005), Mukaidono (2001) and Ross (2004).

Fuzzy logic provides a framework to analyse engineering problems that cannot be characterised with a crisp performance function or model. Being an extension of two and many-valued logics, the main advantage of the fuzzy approach is its capacity to

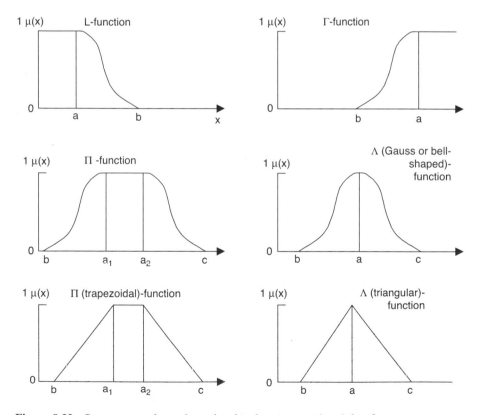

Figure 8.32 *Some commonly used membership functions used to define fuzzy sets*

combine available data, expert knowledge and (subjective) experience in order to mimic real-world conditions more realistically. As an approach to dealing with uncertainties, soft data and signs of varying intensity, it complements other theories such as evidence theory, rough set theory and probability theory. There is still an ongoing controversial discussion between supporters of the fuzzy approach and protagonists of the crisp probabilistic approach. The debate reduces to the question whether fuzziness is just probability in a clever disguise (Bezdek 1994). However, it can be stated that the philosophical and academic controversy, as necessary as it is, clearly has been passed by successful fuzzy applications. Fuzzy mathematics is applied especially in industrial controlling, in medicine and economics and increasingly also in environmental modelling.

Fuzzy logic is also applied to expert systems. An expert system (ES) or knowledge-based systems (KBS) can be seen as a link between specialised human expertise and the end user. An expert system uses knowledge and inference procedures to model human reasoning in a narrow, specialised field (Bichteler 1986). Expert systems exist for a variety of objectives and disciplines and approximate heuristic inference for defined problems. They incorporate domain knowledge and reasoning of human experts and combine it with the computer's processing ability and memory capacity. Based upon queries, an expert system draws conclusions about a certain subject of consultation. The system continues to request data and information until it reaches one or more conclusions.

When data and information are incomplete, vague, or fuzzy, uncertainty management becomes an important function. However, in spite of uncertainties, an expert system should be able to complete inference. Consequently, fuzzy expert systems (FES) have been introduced with rule bases adapted to fuzzy information. Heinrich (2000) uses such a FES to establish the soil assessment fuzzy expert system (SAFES). SAFES is conceptually simple; it consists of an input, a processing and an output stage. During the input stage, the assessor's data are coded to appropriate membership functions. This coding results in confidence values or membership degrees. The inference engine invokes each appropriate rule and combines and aggregates the results of the rules. During the output stage, the conversion of the aggregation back into crisp output values results in soil degradation potentials.

Other applications based on fuzzy logic have been developed. Fuzzy logic provides a powerful tool of exploiting information that would otherwise be discarded as vague or ambiguous. It combines this information with data from real measurements. This qualifies fuzzy logic as an elegant approach to processing information from site investigation work.

Revealing the contamination potential with aerial photos, building permits and fuzzy logic (Heinrich 2000)

The site is a typical harbour terrain with quasi zero relief energy and a ground elevation of about 4 m above sea level. Covering an area of 9 ha, it was used to treat wastewater from ships and other sources. The geological map of 1940 indicates mainly a Holocene deposition with young marine sands less than 2 m thick overlaying marine sands with clay lenses. Major constructional changes affected the whole area: during the creation of the harbour basin in the early 1960s, the ground level was raised about 4 m above sea level. To elevate the terrain, excavated material from the harbour basin may have been used and probably also external filling material was used. This implies the contemporary geology is of anthropogenic character (urbic anthrosols).

The historical analysis revealed that the site hosted three different generations of wastewater treatment plants. A consequence of the environmental laws amendment concerning the contaminant discharge into sewer systems and receiving waters was that the quantity of treated waste increased in the 1980s. These activities subsume the handling and treatment of liquid waste from ships, tank cars, containers, land-storage tanks and industrial plants. The installations had a total annual capacity of treatment of 60 000 tons of liquid waste. The available storage places had a maximum capacity of 1000 barrels with a total volume of 250 m^3.

Investigation on-site was difficult. Access was restricted due to on-going legal procedures. It was therefore decided to evaluate the sites by means of aerial photographs and the available official company documents (statistics, building permits, etc.). The first step was analysis of the processes that led to the degradation of the site, using official records and maps. This analysis included all production, transformation, and deposition processes that had been recorded. Then a sequence of all available aerial photographs was analysed in order to establish an inventory of items characteristic to the site such as tanks, office buildings, processing installations, pipelines, ground discolourations, etc.

In the next step, a logical base of rules was set up to define the theoretical contamination potential of the observed items. Since crisp information was not available, the rules were projected onto fuzzy sets. An example for a rule of this kind is (see Figure 8.33):

IF a large storage tank is observed AND an unusual soil discolouration is spotted next to it OR an unusual soil discolouration is spotted next to a pipeline leading to this storage tank, THEN a high potential of soil contamination is given.

The observed phenomena were described with predefined fuzzy rules and their associated fuzzy membership functions. Both form the core of the expert system that served as an interrogation tool in the following step, a second analysis of the photo series. Based on this analysis, fuzzy contamination potentials were derived for all observed items. The results were stored in the database of a geographical information system (GIS) and visualised with maps of contamination potentials of the observed items. Finally the contamination potentials of all observed items were combined in a continuous map of fuzzy contamination potentials based on a kriging routine (explained below) as shown in the upper map of Figure 8.34.

Later, exploration borings were conducted and samples from a different campaign were evaluated and kriged to derive a second map (lower illustration in Figure 8.34) of contamination potential, this time based on real data. A comparison of the maps reveals that the fuzzy evaluation based on aerial photographs and company records coincides with the contamination map from the field data.

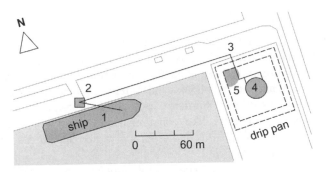

Figure 8.33 *(1) ship; (2) pumping installation; (3) pipeline; (4) tank and (5) surface discoloration i.e. a possible soil contamination*

8.3 Interpretation

The interpretation of information collected during a site investigation is based on the spatial representation of the data. This is usually in the form of maps, profiles and 3D-presentations. Different sets of data may be correlated, combined and processed with geo-information systems to create new data and to characterise and visualise spatial phenomena.

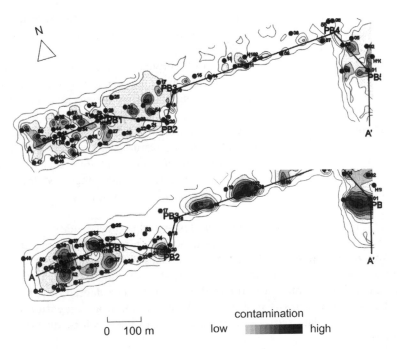

Figure 8.34 *Fuzzy assessment of contamination potential from aerial photographs (upper) and contamination potential from field data (lower); darker grey tones indicate higher contamination potential*

193

8.3.1 Mapping information

In the course of a site investigation campaign, data are acquired in all sorts of forms. Topographic maps depict elevation, land use features, rivers and lakes, etc., historical maps depict the former use of the land, thematic maps depict special features like soil types and ecosystems and aerial photos and satellite scans represent the EM reflectance for a selected spectrum. Furthermore, field measurements provide point information on certain parameters like contamination, salinity or depth to the groundwater table. In addition, records and reports give non-spatial descriptions on phenomena, processes and statistics.

All this information has to be evaluated and harmonised on the *status quo* map. For this task, a general distinction can be made between

- discrete features
- continuous features and
- non-spatial information.

Discrete features are points such as test locations, lines such as boundaries and concrete objects such as buildings and roads. Continuous features are topographic elevations, grey-tone variations on aerial photos or vegetation densities. Non-spatial information helps to interpret and process discrete and continuous features. In order to produce lucid and easy to read maps, discrete and continuous features have to be superimposed in such a way that the phenomena to be depicted are clear.

Certain data that are collected as discrete information may need to be converted into gradual features to be processed further. Typical examples are point information on contamination, soil acidity and depth to the groundwater table. In these cases, a map depicting continuous information has to be produced from a parameter that is only sampled intermittently. The conventional way of achieving this is interpolating between the measuring points to construct contour lines. Contouring and triangulation techniques, although easy to apply, are not efficient in the sense that all information available is utilised. In the 1950s, D G Krige introduced the concept of regionalised variables (Krige 1951). Data describing gradual features are autocorrelated i.e. compared with themselves in order to exploit information on their spatial variation to optimise the contouring procedure. Many researchers including G Matheron in France further developed this method (Matheron 1962, 1963), today referred to as *kriging*.

Consider a boring along which soil samples are taken every metre to measure the clay content z_i. Make a copy of this sequence and place it next to the original one. A comparison of the data from both sequences shows maximal agreement since they are identical (as in Figure 8.35). Now, start shifting the copied sequence down 1 m. This shifting interval is referred to as lag τ. The degree of agreement between the two data series is now lower. Autocorrelation as a measure of the agreement between a data set when comparing it with itself can be quantified as autocovariance:

$$\mathrm{cov}_{z,\tau} = \frac{1}{n_\tau} \sum_{i=1+\tau}^{n} (z_i z_{i-\tau} - \bar{z}_i \bar{z}_{i-\tau})$$

194

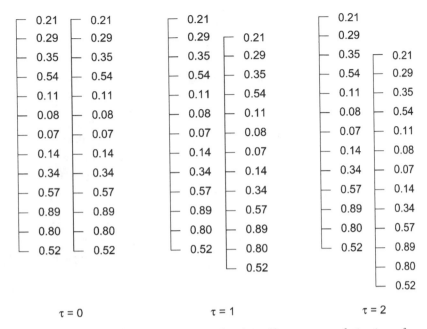

Figure 8.35 *If a series of data is compared with itself, an autocorrelation is performed*

where n_τ is the number of data z of the data sets that are compared. Dividing the autocovariance by the standard variations of the matching series yields the autocorrelation coefficient:

$$r_{z,\tau} = \frac{\mathrm{cov}_{z,\tau}}{s(z)\,s(z_\tau)}; \qquad -1 \le r_{z,\tau} \le 1$$

A diagram with autocorrelation plotted as a function of the lag τ is called an autocorrelogram. The autocorrelation decreases with increasing τ. However, matching of data may recover with increasing lag, indicating a cyclic phenomenon. A cyclic repetition is commonly observed when analysing borehole data or time sequences, corresponding to sedimentary or climate cycles, for example.

Another way of deriving a measure of self-similarity is directly based on the z_i–$z_{i+\tau}$-scatter plot (Isaaks and Srivastava 1989); see Figures 8.36 and 8.37. With the 45°-line defining maximal coincidence, the distances of data points from this line indicate the degree of similarity. In other words, the average distance of a data point from the 45°-line defines a measure of self-similarity. The moment of inertia about the 45°-line provides a measure of this. For a single data point, the squared distance perpendicular to the 45°-line follows from Pythagoras' Law. The average of all points consequently follows from

$$\gamma = \frac{1}{2n_\tau} \sum_{i=1}^{n_\tau} (z_i - z_{i+\tau})^2$$

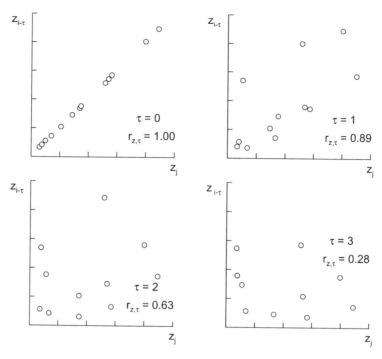

Figure 8.36 *Autocorrelation can be visualised with scatter plots: at zero-lag the data points are aligned along a 45°-line, whereas with greater lag the data scatter increasingly*

where n_τ is the number of data pairs to be compared for every lag τ and γ is the mean squared difference between the original observations z_i and the shifted data $z_{i+\tau}$. γ is referred to as semivariance ('semi' because of the division by 2). A diagram with the semivariance plotted as a function of the lag is called a semivariogram. The

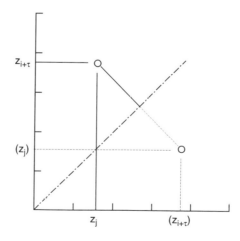

Figure 8.37 *From the scatter plot, the degree of similarity can be deduced directly*

196

semivariance increases with increasing lag, whereas the autocovariance (and consequently the autocorrelation coefficient) decreases. Hence, semivariance and autocovariance (or autocorrelation) are complementary (Figure 8.38).

In semivariance analysis, the mathematical formulation is slightly modified by introducing a lag distance h instead of keeping the autocorrelation lag τ (Davis 2002, for example)

$$\gamma_h = \frac{1}{2n_h} \sum_{i=1}^{n_h} [z(x_i) - z(x_i + h)]^2$$

After a certain distance a plateau is reached at which the semivariance no longer increases, implying there is no longer a correlation to the original data set. This plateau is called the sill and corresponds to the variance of the observations, or the autocovariance at zero lag. The distance to this plateau is referred to as the range, a. Experimental semivariograms can be fitted to model semivariograms to assist further processing. Linear models and non-linear models have been proposed. The failure of the experimental semivariogram to go through the origin is referred to as the nugget-effect, indicating high variability over distances smaller than the sampling interval.

As explained in textbooks on geostatistics, semivariograms can also be derived for data points scattered over a map. Being a measure of self-similarity, semivariograms can thus be exploited to optimise mapping of continuous features from point measurements or, in other words, to estimate unsampled locations y_p from sampled locations y_i in a statistically optimised way. The value at an unsampled location is related to the values at sampled locations. With decreasing distance, the value of a sampled location determines the value of the unsampled location (Figure 8.39). This influence can be expressed with weights w_i that are a function of the distance d_i

$$y_p = \sum_{i=1}^{n} w_i y_i; \qquad w_i = f(d_i)$$

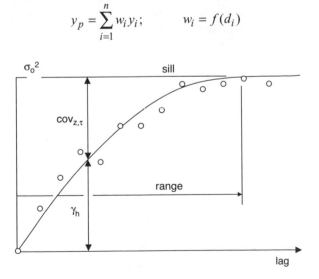

Figure 8.38 *Semivariance and autocovariance are complementary*

197

where y_p is a linear estimator and

$$\sum_{i=1}^{n} w_i \overset{!}{=} 1.0$$

i.e. is true in all cases. Weights may be defined by linear interpolation

$$\sum_{i=1}^{n} w_i = \frac{1/d_1}{\sum\limits_{i=1}^{n} 1/d_i} + \frac{1/d_2}{\sum\limits_{i=1}^{n} 1/d_i} + ... + \frac{1/d_n}{\sum\limits_{i=1}^{n} 1/d_i} \overset{!}{=} 1$$

yielding the estimator

$$y_p = \frac{y_1/d_1}{\sum\limits_{i=1}^{n} 1/d_i} + \frac{y_2/d_2}{\sum\limits_{i=1}^{n} 1/d_i} + ... \frac{y_n/d_n}{\sum\limits_{i=1}^{n} 1/d_i}$$

However, weight factors can be defined in many different ways. As well as simple inverse distance functions $1/d_i$, powered functions $1/(d_i)^n$ or even more complex approaches are feasible.

The optimal set of weight factors can only be derived if, besides the distance d_i, a second parameter is introduced. For a kriging routine this is done by considering the self-similarity of all data points. As explained previously, self-similarity may be expressed as semivariance. Including the semivariance to optimise contouring constitutes the basic idea of kriging. Based on this approach, best estimates at points where no data were acquired can be made from a given set of measured points. In addition, kriging enables us to quantify the estimation error (as explained in the case study). The combination of these two aspects makes kriging a powerful tool to enable the visualisation of gradual features.

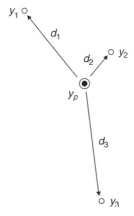

Figure 8.39 *Unsampled location y_p with three neighbouring sampling points y_1, y_2 and y_3*

A demonstration of kiging (Davis 2002, Clark 1979)

In order to illustrate how kriging works a simplified example is introduced, where three sampling points serve to estimate an unsampled location (Figure 8.40), a process also referred to as punctual kriging. Suppose that a semivariogram for these three sampling points has been established. To deduce the optimal set of weighting factors, the semivariances are associated with the weighting factors in three simultaneous equations:

$$w_1\gamma(d_{11}) + w_2\gamma(d_{12}) + w_3\gamma(d_{13}) = \gamma(d_{1p})$$
$$w_1\gamma(d_{21}) + w_2\gamma(d_{22}) + w_3\gamma(d_{23}) = \gamma(d_{2p})$$
$$w_1\gamma(d_{31}) + w_2\gamma(d_{32}) + w_3\gamma(d_{33}) = \gamma(d_{3p})$$

where w_1, w_2 and w_3 are the weighting factors and $\gamma(d)$ the semivariances for the distances d_{ij} between control points i and j, as illustrated in Figure 8.40. $\gamma(d_{13})$ is, for example, the semivariance for the distance between measuring point 1 and 3. $\gamma(d_{1p})$ is the semivariance for the distance between measuring point 1 and the location p where the estimate is to be made. A fourth equation defines the sum of the weighting factors as 1.0:

$$w_1 + w_2 + w_3 = 1.0$$

Thus, four equations are derived with only three unknown variables. We take advantage of this extra degree of freedom and introduce the Lagrange multiplier λ allowing us to minimise the estimation error of the unsampled location p

$$w_1\gamma(d_{11}) + w_2\gamma(d_{12}) + w_3\gamma(d_{13}) + \lambda = \gamma(d_{1p})$$
$$w_1\gamma(d_{21}) + w_2\gamma(d_{22}) + w_3\gamma(d_{23}) + \lambda = \gamma(d_{2p})$$
$$w_1\gamma(d_{31}) + w_2\gamma(d_{32}) + w_3\gamma(d_{33}) + \lambda = \gamma(d_{3p})$$
$$w_1 + w_2 + w_3 + 0.0 = 1.0$$

or, in matrix form

$$\begin{vmatrix} \gamma(d_{11}) & \gamma(d_{12}) & \gamma(d_{13}) & 1 \\ \gamma(d_{21}) & \gamma(d_{22}) & \gamma(d_{23}) & 1 \\ \gamma(d_{31}) & \gamma(d_{32}) & \gamma(d_{33}) & 1 \\ 1 & 1 & 1 & 0 \end{vmatrix} \cdot \begin{vmatrix} w_1 \\ w_2 \\ w_3 \\ \lambda \end{vmatrix} = \begin{vmatrix} \gamma(d_{1p}) \\ \gamma(d_{2p}) \\ \gamma(d_{3p}) \\ \gamma(d_{1p}) \end{vmatrix}$$

that is

$$A \cdot \vec{W} = \vec{B}$$

where the elements of matrix A and vector B are extracted from the semivariograms established for the site under investigation (Figure 8.40). Solving this set of equations yields the estimator for the unsampled location

$$y_p = w_1 y_1 + w_2 y_2 + w_3 y_3$$

which has an estimation variance of

$$s_e^2 = w_1 \gamma(h_{1p}) + w_2 \gamma(h_{2p}) + w_3 \gamma(h_{3p}) + \lambda$$

and thus an estimation error of

$$s_e = \sqrt{s_e^2}$$

This simple example illustrates that kriging yields both the estimator and the error associated with this estimation, an aspect that distinguishes this approach from all other contouring routines.

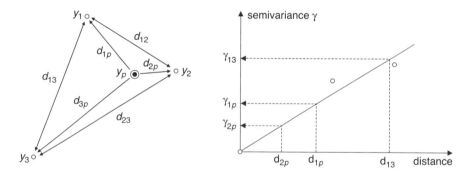

Figure 8.40 *Map with measuring points y_1, y_2 and y_3 and point to be estimated y_p. Semivariogram with semivariances γ taken for difference distances*

Mapping organic contaminants on Moltke Chemical Plant and Mine (Genske 2003)

The former mine and coking plant Graf Moltke 3/4 is located in the northwest of the Ruhr District close to the city of Essen, Germany, and covers an area of about 230 000 m². In 1873, the first shaft was sunk, followed by another 3 during the 30 years that followed. A coking plant was built in 1903–04 (Figure 8.41), followed by benzole and ammonia factories. Over the following 50 years, additional industries were established, turning the site into a multi-use industrial complex. Coal production went up to 1 Mton annually in the 1960s, until the mine was closed in 1971.

In the years that followed, Graf Moltke became an industrial wasteland with all the typical negative attributes. Positive points were, however, the existing and still intact infrastructure of the surroundings and the immediate proximity to the A2 Autobahn, one of the most frequented German highways and an important east-west traffic connection, especially in light of German reunification. In the late 1980s, it was therefore decided to remediate the site as a project of the *Internationale Bauausstellung IBA Emscherpark*, and to convert it to a high-quality industrial park. The European Community provided the project with appropriate funding via the European Fund for Regional Development (EFRE), so that only about 50 % of the remediation costs had to be met by the owners of the former mining site.

In the first phase of the remediation process, a multitemporal analysis was carried out. Based on this evaluation the field investigation program was planned. The samples taken were analysed in the laboratory and the type of contamination specified. The sampling points, which were distributed irregularly over the site (Figure 8.42), were interpreted as regionalised variables and a block kriging routine was applied (to segments or blocks, not to individual points). In addition to the map depicting contaminated sectors (Figure 8.43), the relative prediction error was quantified (Figure 8.44). The relative prediction error refers to the ratio of the absolute error over the contamination observed. Based on the error map, the field investigation program was further optimised: sectors with both high contamination and large error indicated that additional investigation was necessary.

Figure 8.41 *Graf Moltke Mine in 1912*

8.3.2 Associating information

It has been mentioned in previous chapters that data can be superimposed and correlated to gain spatial information on certain phenomena. In the 1950s and 60s the idea of data layers that can be superimposed was pursued at different laboratories, including the Massachusetts Institute of Technology (MIT), the ETH Zürich and the Harvard Laboratory for Computer Graphics and Spatial Analysis. Roger Tomlinson developed the first functional geographical information system (GIS) in Ottawa, Canada, in an

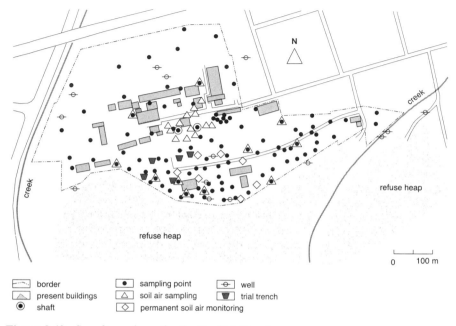

border	● sampling point	well
present buildings	△ soil air sampling	trial trench
⦿ shaft	◇ permanent soil air monitoring	

Figure 8.42 *Sampling scheme for the Graf Moltke site*

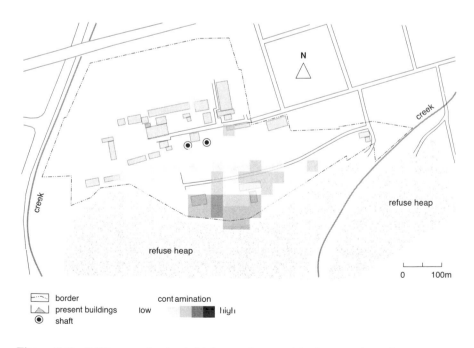

border	contamination	
present buildings	low ▬▬▬ high	
⦿ shaft		

Figure 8.43 *PAH contamination is highest at the site of the former coking plant*

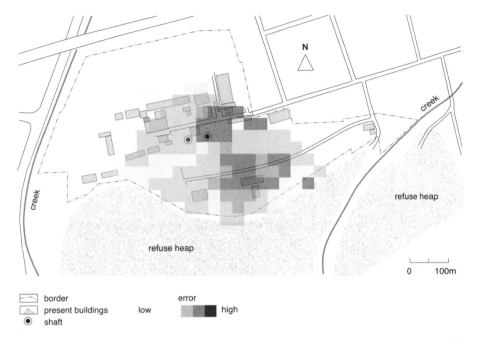

border
present buildings
shaft
error
low   high

Figure 8.44 *Relative spatial error of measured contamination, suggesting where further samples have to be taken*

attempt to visualise the capacity of the terrain for selected land use such as agriculture, forestry, wildlife etc. He also organised the first GIS Symposium in 1970, during which the term 'Geographic Information System' was coined. This marked the first phase of GIS development, the time of GIS pioneers (Bartelme 2005). A phase of municipalities aimed at utilising GIS to ease administration work followed, then a GIS market developed with powerful GISs being introduced. Thereafter, users developed specialised GIS applications based on commercial GIS frames, which eventually led to open-source GIS software enabling individual users to use GIS as a tool to solve their specific problem (Neteler and Mitasova 2004, Kropla 2005, Longley *et al.* 2005).

Today, GISs are used as a common tool for creating, associating and processing spatial data and information attributed to them. There are two types of data that are processed in a GIS:

- Vector data: discrete information such as points, lines or areas representing e.g. wells, boundaries and man-made structures (for example, Figure 8.45) Every vector object has its defined place on the map as well as certain attributes. For a dismantled building on a derelict industrial site, for example, the time of construction, the function, the time of dismantling, the owner, etc. is stored.
- Raster data: continuous information such as aerial photos, satellite scans and grids representing e.g. elevations or rainfall. Every grid point or pixel has its place in the grid as well as a unique attribute e.g. elevation, grey tone, wavelengths, etc. Pixels can be combined to form larger units to improve the presentation and reduce storage space.

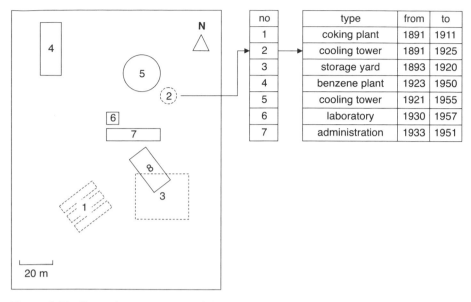

Figure 8.45 *Vector layer in a GIS and the database to store attribute data*

Attributes are thus items of information associated with vector objects, or raster data that can be analysed, superimposed and modelled. Vector objects may have a multitude of attributes whereas raster data are only associated with single items of information. Vector and raster data are stored in databases for further processing.

The development of a GIS involves:

- Layers of information are established that provide either vector data or raster data. The data are associated with attributes.
- The different layers are conditioned and refined to remove errors and to adjust the scale. All entities have to be located accurately to the given coordinate system, a process referred to as georeferencing. Automated procedures have been developed to aid this process.
- The different layers with their attributes are simplified, generalised, stored for further analysis and superimposed to extract the information desired.

Data can be captured from all kinds of formats. Conventional maps are scanned to produce both raster data for gradually changing features and vector data for discrete objects. Stereopairs of aerial photos are automatically analysed to produce topographic maps. Satellite scans are imported as raster data to form individual layers of information.

If necessary, a transformation between the two data formats can be performed. For instance, vector data in the form of point information such as rainfall measurement can be used to create an intensity map by means of kriging procedures as explained above. These maps can than be interpreted as raster data.

The following tasks can be performed by means of a GIS:

- Visualising selected information e.g. soil type, rain intensity or slope inclination.

- Visualising spatial relations between objects of the same type e.g. distances between well points or objects of different kinds e.g. distances between buildings and streets.
- Combining objects with the same attributes or with selected attributes e.g. soil types of low permeability or plant species, indicating contamination.
- Superimposing information to extract areas of intersection.

The final point listed above highlights the usefulness of a GIS. For example, if searching for a suitable location for a landfill, a layer indicating low ground permeability may be superimposed with a layer of the site indicating where building permits will be granted, a layer indicating groundwater protection zones and a layer indicating the minimal distances to buildings (which should not be too close to the landfill). GIS enables the user to compose and superimpose specific layers to filter out the information relevant to a particular project.

Another example is the remediation of a brownfield. During the desk study phase, historic maps and aerial photos are compared to geological, soil and hydrogeological maps etc. to deduce information on possible ground contamination. This superposition of information is quite a demanding process, since the complexity of an industrial installation has to be studied beforehand, including processes of mass flows relevant to the different generations of utilisation and accidents which occurred while the site was in use.

Another powerful facility of GIS is the option to

- analyse and model the attributes database to produce additional thematic layers.

For example, the normalized difference vegetation index (NDVI) can be mapped by processing two bandwidths of Landsat imagery. Erosion potentials can be mapped by processing information on erodibility, erosivity, slope inclination, land cover and land management. In the same way, the migration of contaminants in an aquifer can be modelled when investigating the potential of reusing degraded land. The SAVES project described previously is a typical GIS modelling tool based on fuzzy logic. It uses an interactive knowledge base to deduce contamination potentials from aerial photos and building permits. In addition, digital surfaces can be produced and animated in three dimensions. In certain cases, profiles can be drawn to explain certain processes like the migration of pollutants or the development of subsidence.

Reutilisation potentials of derelict land in the State of Thuringia, Germany

An integral part of sustainable land management is the recycling of used land. Land recycling releases the pressure on natural land and eliminates possible contamination sources. A key problem of land recycling is assessing the potentials for different reuse options. These options include reuse as a residential area (individual homes and apartment buildings), commercial use (crafts and trade, production and industry, services and offices) as well as recreation and re-greening. In order to provide a simple and robust tool to assess possible reuse options, an open-source GIS has been developed, based on the following information (**Ruff *et al.* 2006**):

- topographic maps
- publicly available land-registers
- aerial photos
- thematic maps (e.g. geological maps, hydrogeological maps, soil maps, water protection zones, parks, areas prone to flooding, registers of contaminated sites, flora-fauna-habitat-compensation sites, etc.)

Most of this material is available in digital form. In addition, archives can be screened for relevant information on the state of degradation of the ground. For a number of selected communities in the State of Thuringia, this information was stored in a GIS on individual layers with metadata giving abbreviated information for quick screening.

For the GIS developed, the objects of importance are the sites which lie derelict. They constitute the pool of land resources to be recycled. Derelict sites are characterised with attributes like address, size, owner, etc., then all sites are stored in the land management tool ResDbase. Specific operations are performed in order to reveal their reuse potentials. This not only includes the superposition with thematic layers as described previously, but also data modelling e.g. the calculation of the distance between the site and the nearest Autobahn and public transport options. Based on the data available and their modelling, ResDbase derives ratings for the different options of future use, with 100 being the most favourable reuse option and 0 being a reuse of no potential. ResDbase permits the visualisation of reutilisation potentials for derelict site collectives. It is possible to visualise the re-utilisation potential of apartment buildings, for example, for an entire community with areas of high, low and intermediate potential highlighted in different colours. This makes ResDbase a valuable tool when compiling an inventory of the regional resources of derelict sites, and facilitates the decision making process for parties involved in sustainable management of the resource land.

9 Remediation

After discussing natural ground conditions and the way they are degraded as well as strategies to investigate the degree of degradation, measures of remediation are introduced. As in the preceding chapters, a distinction is made between erosion, chemical degradation and physical degradation. The remediation measures introduced are illustrated with examples and case files.

9.1 The man who planted trees

Sometimes it takes little to change one's point of view. A few years ago, I read Jean Giono's (1990) *L'homme qui plantait des arbres* (*The man who planted trees*). The story describes the life of Elzéard Bouffier, who decided to revitalise an abandoned and devastated region where the Alps penetrate the French Provence. The once populated region was devastated, the houses abandoned, the wells dried out. The land was almost barren, with fierce winds ravaging the plains that rise to about 1300 m above sea level. Monsieur Bouffier, who guarded his flock of sheep in this monotonous mountain range, spent his evenings sorting acorns, the best of which he would use the next day to plant oak trees. Every day, he planted a hundred oak trees. Only one of ten acorns would actually start growing. Later, he also started planting birches and other trees. He did not notice that there was a Great War, the First World War. The area was too remote to be affected. He concentrated on his task of planting trees and in order to make his work more efficient he sold most of his sheep since they threatened the young trees and kept beehives instead. The Second World War went by just as the first one did, while he continued planting trees. By 1945, the once deserted terrain had transformed into a vast forest. The creeks returned, just as the flowers and the wildlife that distinguish a mountain forest. The sound of running water and the summer wind caressing the trees had returned. Soon after, Elzéard Bouffier died peacefully almost 90 years old at the hospice of Banon. His work remains a remarkable example of what a single individual can do in his effort to restore nature.

In 2004, Wangari Maathai became the first woman from Africa to receive the Nobel Peace Prize—for planting trees. With her Green Belt Movement she mobilised poor woman to plant 30 million trees to halt deforestation and desertification. The Green Belt Movement paved the way for rehabilitating our environment on a grass roots level while fighting poverty and corruption through education and family planning. In a press release the Nobel Committee stated

"...peace on Earth depends on our ability to secure our living environment. Maathai stands at the front of the fight to promote ecologically viable social, economic and cultural development in Kenya and in Africa. [...] She thinks globally and acts locally."

9.2 Erosion

According to the GLASOD-assessment (Oldeman *et al.* 1991), about 1633 Mha are degraded globally by erosion. This represents 84 % of all human-induced soil degradation. Two-thirds of this area is eroded by water and one-third by wind, 37 % slightly, 48 % moderately and 15 % strongly to extremely degraded.

Erosion consumes the solum and thus reduces the soil profile. In extreme cases the parent material of the C-horizon is denuded. This is accompanied by a reduction in water retention capacity, a loss of filter, buffer and recharge functions and a destabilisation of natural terrain and man-made constructions. As well as these on-site impacts, erosion causes off-site effects such as the silting-up of watercourses and reservoirs and their contamination with chemicals, an increase in flood frequency and a loss in biodiversity. In addition, erosion is believed to play a key role in the deterioration of our climate.

Erosion may be precluded with measures of prevention. Once erosion is already taking place, remediation measures become necessary.

9.2.1 Measures of prevention

Due to the importance of the erosion problem, extensive research has been carried out aiming at measures of preventing erosion. A frequently discussed approach is defining a soil loss tolerance for the area under investigation. Soil loss tolerance refers to the amount of soil that can be lost without the soil losing its capacity to produce food and fibres. Farmers tend to presume a relatively high soil loss tolerance, since they are interested in keeping crop yields high. Ecologists, on the other hand, tend to assume a much lower soil loss tolerance, since they take into consideration the ecological functions of the soil. A reasonable approach would be fixing soil loss tolerance at the rate soil naturally forms. This again is difficult, since soil formation rates vary globally from 0.01 to 7.7 mm yr^{-1} with an average of about 0.1 mm yr^{-1}. Taking off-site effects into account produces another acceptable annual soil loss tolerance of about 1 ton ha^{-1}, the equivalent of 0.1 mm yr^{-1} (Morgan 2005).

From a practical point of view, maintaining vegetation cover appears to be the most efficient way of preventing erosion. An ideal cover is provided by woodland that protects the soil from the impacts of water and wind erosion. If woodland is exploited for timber then clear-cutting should be avoided, instead patch-cutting or selective felling of mature trees exercised instead. When felling trees, damage to younger trees and other plants should be minimised. The loss of trees should be offset by reforestation measures.

On cultivated land, sustainable farming techniques should be implemented. Crops should be planted in such a way that they rapidly cover the soil and reinforce it with their roots throughout the windy and rainy season. When seeding crops, timing is thus important to optimise the shielding effect of the developing canopy. Crop rotation reduces the loss of topsoil, especially when plants vulnerable to erosion such as maize, sugar beet or cassava are cultivated. Figure 9.1 illustrates, for instance, that maize-wheat-clover rotation would reduce a natural soil profile (land under grass) by only 30 % whereas a maize monoculture would reduce it by over half. Cover crops that grow quickly and reinforce the soil with their roots control erosion on idle fields and underneath trees and vines. With strip cropping, along the contour lines or perpendicular to the wind direction, runoff and dust can be trapped on erosion resistant strips. Riparian buffer strips protect rivers and lakes from eroded soil and agrochemicals. Leaving crop residues on the field after harvesting and mulching the fields considerably reduces erosion. The success of mulching can be expressed with the mulch factor MF, that is, the ratio of soil loss with a mulch to that without. In conservation agriculture as promoted by the FAO, techniques to reduce erosion are harmonised with a more efficient use of soil, water and biological resources while respecting traditional ways of land management (FAO 2000).

On grazing land, overgrazing inevitably leads to a loss in vegetation cover and thus to extensive erosion problems. Morgan (2005) states that the vegetation-erosion-grazing interaction appears to be a complex process with many factors to be accounted for including "soil fertility, loss of soil nutrients, production of litter, the palatability and digestive value to stock of the different species in the plant community and the ability of the different species to survive under changing grazing, moisture and nutrient conditions". This interaction has not been fully understood so far. A sustainable practice that minimises erosion is rotational grazing in a nomadic way as practised traditionally in regions not dominated by the commercial overexploitation of rangeland.

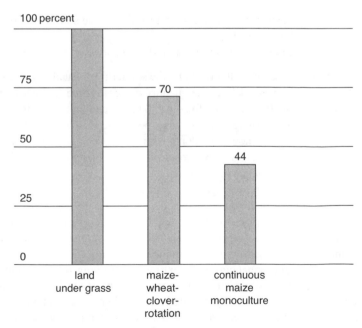

Figure 9.1 *Reduction of topsoil thickness after 100 years for land under maize-wheat-clover rotation and continuous maize monoculture, compared to a permanent grass cover for a 4° slope of silt-loam soil (after Gantzer et al. 1990)*

On urban land, construction works and the accompanying denudation of soils lead to considerable erosion problems. In order to minimise erosion, soil exposure needs to be scheduled with seasons of low precipitation and moderate winds. Bare construction roads should be covered with gravel and possibly reinforced with geotextiles. Silt fences should be provided to prevent soil being washed off construction sites into rivers and lakes. At the end of construction work, exposed soil has to be re-vegetated as quickly as possible. On derelict land and industrial fallows (brownfields) a stable vegetation cover should be maintained, not only to restrict water and wind erosion but also to reduce contaminated runoff and dusts.

Natural habitats are hardly affected by erosion as long as people are kept from entering erosion prone areas. Seeding erosion resistant plants along trails for hikers and horses precludes erosion problems. Local plants should be utilised to avoid disturbing the given ecosystem. Sport activities such as mountain biking and off-road riding may increase the risk of erosion, especially if hiking and riding paths are reused repeatedly. The question should be raised whether the promotion of off-road vehicles in developed countries rather serves the profits of industry than the notion of individual freedom.

9.2.2 Measures of remediation

If erosion is already underway, measures of remediation have to be taken. Such measures aim at (Morgan 2005)

- re-establishing and maintaining the plant cover i.e. phytostabilisation of the surface
- reducing the erodibility i.e. the susceptibility of the soil to erosion
- improving the soil profile
- introducing engineering measures of stabilisation and drainage.

Phytostabilisation begins with increasing the soil fertility, usually by applying organic matter. The isohumic factor defines the quantity of humus produced per unit of organic matter and ranges from 0.20 for plant foliage to 0.65 for coniferous tree litter and 0.85 for peat moss. Other means of fertilisation may be applied such as manure or artificial fertilisers. However, this should be done with care, since over-application leads to contamination and eutrophication problems. In addition, conditions have to be maintained that allow plants to grow, which includes monitoring the soil moisture, the pH, the salinity and other factors controlling re-vegetation measures. The new vegetation cover should have an elevated level of biodiversity since this permits mechanisms of self-defence against pests and invasive plants. Reforestation leads to an optimal reduction of erosion.

Erodibility, an inherent soil property, varies with aggregate stability, soil texture, porosity and permeability. Measures to reduce erodibility therefore aim at increasing aggregate stability and enhancing soil texture by applying organic matter and clay minerals that bind the soil and make it more resistant to the forces of erosion. Soil conditioners such as organic by-products, polyvalent salts and various synthetic polymers may, although expensive, stabilise the soil temporarily until a vegetation cover has established.

The soil profile can be improved by introducing barriers both along the contour lines to reduce the speed of runoff, and perpendicular to the main wind direction to reduce the speed of surface winds. Contour tillage is a well-established method to reduce fluvial erosion on cultivated land. In addition, contour bunds may be placed to block runoff. They may be constructed as small earth banks of 1–2 m width or as stone bunds, with smaller stones placed upslope to trap sediments. In combination with buffer strips as described above, contour tillage and contour bunds considerably reduce erosion. Gullies may be stabilised by installing small dams perpendicular to the gully direction. These dams serve to retain the eroded soil and to prevent gullies deepening with every rainstorm event. They may be combined to form larger sedimentation ponds in order to maintain a maximum of the eroded soil on the site. The ancient technique of terracing was especially designed to retain as much soil as possible to increase crop yields. Morgan (2005) compares the efficiency of different types of terraces including diversion, bench and retention terraces.

Draining slopes is not only a common strategy to reduce runoff and thus erosion, it is also recommended to increase the overall stability of the slope. Stopping water from infiltrating the slope lowers the water table within the slope. As a consequence, a smaller portion of the slope is submerged and subjected to pore water pressures that decrease the shear strength of the soil. In addition, cohesive layers are kept dry and thus maintain their shear strength (Chapter 4). In order to stop runoff from damaging the vegetation cover and subsequently entering the slope in an uncontrollable way, waterways lead the runoff to suitable disposal points. Grass waterways are considered the cheapest and most efficient way to safely drain sloping terrain. For steeper slopes, stone lined channels become necessary.

Techniques of bioengineering take advantage of living plants that are fixed to the slope in rows perpendicular to runoff direction. Fascines (or wattles) are cigar-shaped bundles of six to eight easy-rooting life cuttings (usually *Salix*, *Leucaena*, *Baccharis* and *Tamarix*) that are placed in shallow trenches along the contours at 4 m intervals on slopes less than 30 degrees (Morgan 2005), as in Figure 9.2. Rooting bushes and hedges may also be introduced directly into berms excavated along the contours (Schiechtl 1991), as depicted in Figure 9.3. Steeper slopes may be stabilised by means of wooden grids with rooting plants attached to the slope (Figure 9.4). Gabions i.e.

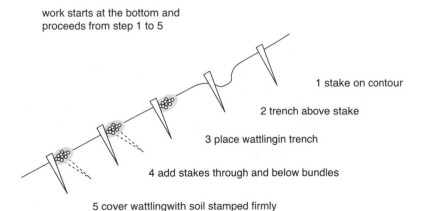

work starts at the bottom and proceeds from step 1 to 5

1 stake on contour

2 trench above stake

3 place wattlingin trench

4 add stakes through and below bundles

5 cover wattlingwith soil stamped firmly

Figure 9.2 *Placing fascines to stop erosion (after Gray and Leiser 1982)*

211

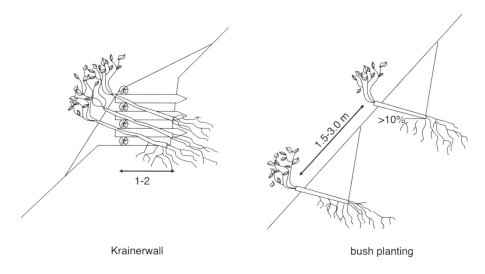

Krainerwall   bush planting

Figure 9.3 *Krainer-walls and bush planting (after Schiechtl 1991)*

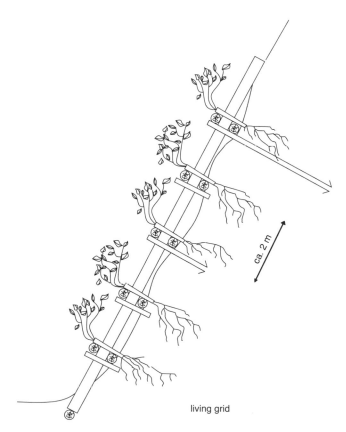

living grid

Figure 9.4 *Living grids stabilise steep slopes (after Schiechtl 1991)*

Figure 9.5 *Gabions (not vegetated) in Nordhausen, Germany*

rectangular steel wire-mesh baskets packed with stones are particularly useful in stabilising the foot of a slope (Figure 9.5). Since they are permeable, they also serve to drain the slope. They may deform without losing stability while indicating which part of the slope is in motion. Hand-packed gabions are much more robust than steel-wire baskets that are simply filled-up with gravel. It is possible to combine gabions with re-greening measures, as shown in Figure 9.6. Slopes may also be reinforced

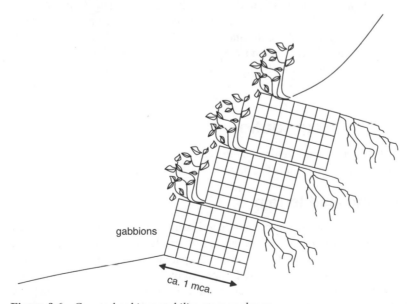

gabbions

ca. 1 mca.

Figure 9.6 *Greened gabions stabilise steeper slopes*

with geotextiles, a technique known as *terre armée* (Vidal 1966). Reinforced soil may be vegetated to enhance both appearance and biological function. In fact, bioengineering has become an attractive application that applies when soft remediation techniques in line with ecological standards are preferred.

In order to reduce the impact of wind erosion, living windbreaks or shelterbelts may be installed. They are placed at right angles to the wind direction and considerably reduce wind velocities both upwind and downwind of the barrier. Shelterbelts may consist of hedges or trees or a combination of both. A mixture of different species enhances the resistance to pests and diseases. Local plants should be used to avoid uncontrollable invasion of exotic species. Although windbreaks can also be quickly erected with dead material (for example plastic fences), living windbreaks are always preferable because of their potential to contribute to local biodiversity and their value to wildlife.

Erosion control at Grand Goulet, French Alps (Rey and Nicot 2004)

Along the alpine scenic road that traverses the Grand Goulet natural site in the French Alps, rock fall has caused considerable damage especially upslope of the road. In order to control the increased erosion risk due to the damage done, remediation measures were taken, focusing on the stabilisation of the vegetation cover. In a first step, the geological conditions and the types of soil, the climatic conditions and the vegetation dynamics were analysed. Thereafter, it was decided to establish a 550 m palisade along the contour lines at distances of 3–5 m over an area of 5 ha. The palisades have a height of about half a metre and are made of stakes backed by cuttings of willow (*Salix*, especially *Salix purpurea* and *Salix incana*), as shown in Figure 9.7. On top of the palisades and between them plants typical for the region were planted including *Buxus sempervirens*, *Acer campestre* and *Sorbus aria*. In addition, the slope was mulched with local plant residues.

Just two months after the rehabilitation measures began, the seeded plants, bushes and grass developed into a stable vegetation cover. The cuttings, especially *Salix purpurea*, developed both roots and leaves. Neither excessive runoff nor gullying had been observed. Research now continues to determine the most effective distance between the palisades and which plant species are most efficient in the rehabilitation work.

9.3 Chemical degradation

As discussed in Chapter 6, chemical degradation comprises contamination, acidification, salinization and solidification. In this Chapter measures of remedy are introduced.

9.3.1 Contaminated terrain

Worldwide, about 1.1 % of human induced soil degradation is caused by contamination (Oldeman *et al.* 1991). The entire soil profile may be affected, from the biochemically active L-, A-, and B-horizons to the parent material of the C-horizon. Both surface water and groundwater may be spoiled by contamination. The mobility of the contaminants depends on their physical and chemical properties, how they are released into the environment and geological boundary conditions (Chapter 6).

| (a) | (b) |

Figure 9.7 *Rehabilitation of a slope prone to erosion along the Grand Goulet Road in the French Alps, (a) immediately after the stabilisation work and (b) two month later (photographs courtesy of Freddy Rey and François-Xavier Nicot)*

A distinction is made between

- local contamination at well-defined contamination hot spots and
- diffuse contamination spread over vast areas.

Certain scenarios are typical for local contamination like spills and accident. Others present diffuse spreading as, for example, the application of agrochemicals to farmland or the fall-out of atmospheric contaminants. A contamination hot spot, although initially confined to a restricted area, may subsequently start spreading, for example, by mixing with the groundwater.

There are many strategies to respond to the degradation of soil by contamination. They include measures of prevention and of remediation.

9.3.1.1 Measures of prevention

Since society became aware of the noxious effects of contaminants, notably after Rachel Carson published *Silent Spring*, efforts have been made to reduce the amount of toxic products and to minimise their release into the environment. Unleaded gasoline has been introduced, just as biodegradable detergents and eco-friendly fertilisers.

In addition, eco-standards have been established to improve production routines, transport and handling of substances with a high contamination potential. The introduction of ISO 14001 *Environmental Management Systems* appears to be a huge step towards a reduction of resource consumption and waste production (Edwards 2004, Whitelaw 2004). As a certification standard, ISO 14001 aims at a continual revision of the activities of an organisation or an industry in order to minimise harmful effects on the environment while at the same time enhancing production processes.

Nevertheless, large amounts of contaminants are still released and many contaminants that were released before environmental standards were improved can still be traced. In addition, serious accidents still happen although measures of prevention have been introduced and awareness has been raised.

9.3.1.2 Measures of remediation

A variety of remediation options have been developed. They may be grouped into

- passive measures and
- active measures.

Passive measures are designed to block contamination pathways between the source and the goods to be protected. Active measures, on the other hand, eliminate or extract the contamination source.

In civil engineering terms one may also distinguish between

- *ex-situ* methods and
- *in situ* methods,

i.e. techniques to extract and clean the soil and to clean the soil without excavating it, respectively. Furthermore

- on-site and
- off-site measures

are distinguished i.e. cleaning measures that are carried out at the place of contamination and at soil cleaning facilities away from the contaminated site, respectively. In certain cases, contaminated soil may be too expensive to treat and is thus deposited at landfills, sometimes on-site, but in most cases off-site.

The most common remediation techniques are presented for the following:

- excavation and dumping
- excavation and treatment, including soil washing, thermal treatment and biological treatment
- isolation
- stabilisation
- hydraulic measures
- bioremediation
- phytoremediation.

In addition, the concept of

- natural attenuation

is introduced. The suitability of these methods depends on many factors including geological conditions, resource efficiency, environmental compatibility, acceptance by the administration and the public, the time frame set for the remediation work and the available budget (Genske 2003).

Excavation

Excavating the contamination source appears to be a straightforward solution to eliminate a contamination hot spot. It is considered a fast, efficient, and definitive remediation method. However, extracting a contamination hot spot consumes many resources since equipment and energy have to be provided to excavate and handle the polluted soil. Clean soil has to be furnished to refill the excavation. In addition, the excavated material has to be taken care of. Health and safety standards have to be observed, emissions of dust and odours have to be controlled and noise minimised.

After excavation, the soil is separated into high, medium and low contaminated material according to the local regulations. Contaminated soil may still be used, for example, as backfill or to cover contaminated terrain. Highly contaminated soil may be either

- dumped at landfills or
- treated to extract or destroy the contaminants.

Landfilling is not in line with the notion of sustainability as this doesn't solve the contamination problem; clean-up work is handed over to the next generations. In addition, since landfills consume space, they contribute to the consumption of our natural environment. Moreover, landfills have to be maintained and monitored and therefore cost money.

Remediation of Prosper III (Genske 2003).

Prosper III was established in 1906 as a deep coal mine close to the city of Bottrop in the German Ruhr District, depicted in Figure 9.8. Most of the terrain was occupied by a coking plant and several chemical plants, refining coal and producing by-products. Because of the decline of the coal industry Prosper III closed down in 1986, leaving 29 ha of industrial wasteland partly identified as highly contaminated. Since Prosper III is located in the very centre of Bottrop it was decided to remediate the abandoned site and to revive industrial activities. Guided by the idea of 'Working in the Park' a residential area was also planned. However, no public funding was made available, forcing the owners of the former mine to finance the remediation project at their own expense.

The geology of the Prosper site is characterised by three strata. The upper consists of about 1–3 m of loose debris and artificial filling (*urbic anthrosol*). Foundation fragments of the former facilities were also found in this upper layer, creating extremely inhomogeneous ground conditions. Undisturbed Quaternary sediments, mainly silty sands, were found beneath to depths of about 16–20 m. The bedrock was specified as fractured Cretaceous marls with the upper metres weathered to low permeable soft rock. A continuous groundwater table was identified at about 5 m below the surface. The field campaign revealed that part of the manmade fillings and the upper Quaternary sediments were contaminated with different contaminants from on-site production processes as well as storage and handling of noxious substances. No significant contamination was discovered in the bedrock.

In order to remediate the Prosper III site it was decided to excavate certain sectors to a depth of about 2 m and to replace the excavated ground with clean building soil. Contaminated soil and foundation fragments were heaped at the centre of the site identified previously as highly contaminated, as depicted in Figure 9.9a. This made the movement of about 180 000 m^3 of soil and foundation fragments necessary (Figure 9.12b). An independent laboratory continuously monitored the grade of contamination of the excavated soil. Once a certain contamination threshold was exceeded the soil had to be dumped on a special waste site at high costs. Fortunately this became necessary only occasionally so that the overall budget was not exceeded.

The landscaped heap was covered with a drain and seal system to avoid immediate contact with polluted soil and to prevent infiltration of rainwater (which would cause a migration of the pollutants from above the groundwater table into the saturated zone). Furthermore, a number of observation wells were installed that can be converted into remediation wells in case the groundwater requires further treatment. The landscaped hill was then converted into a park, surrounded by both residential and business sectors, as depicted in Figure 9.9c.

Figure 9.8 *Historical view of Prosper III*

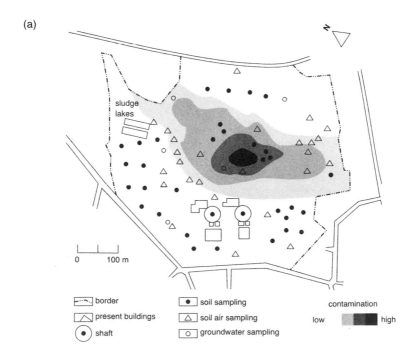

(b)

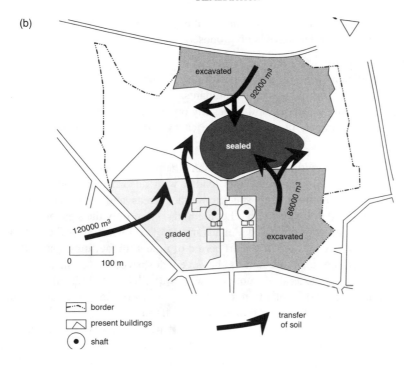

(c)

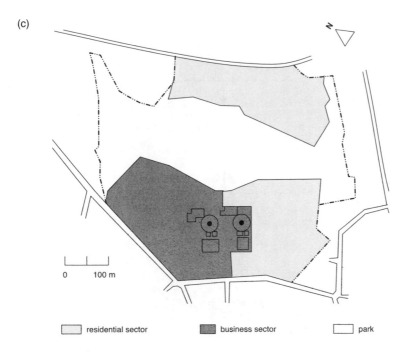

Figure 9.9 *(a) Contamination, (b) mass flows and (c) redevelopment planning at the Prosper site*

A better solution than dumping contaminated excavation in landfills is treating it. The following ex-situ options have been proposed:

- Soil washing: mixing and agitating contaminated soil with water in order to detach pollutants from the soil particles (see Figure 9.10). This option applies to both organic and inorganic contaminants.
- Thermal treatment: heating soil to destroy organic pollutants.
- Biological treatment: biodegrading the excavated soil to eliminate organic contaminants.

During soil washing, additives like surfactants help to increase the solubility of hydrophobic components. Clays and silts loaded with adsorbed contaminants are difficult to wash due to their high specific surface areas and must therefore be separated and deposited or treated with other methods. Thus, only the sand and gravel fractions can be treated and re-utilised. An advantage of soil washing is the fact that contaminated soil can be cleaned in a short period of time with facilities installed on-site. The natural structure of the soil is, however, destroyed — as in all *ex-situ* technologies — and a considerable amount of wastewater has to be treated.

Soil washing was first applied in the 1980s; for example, to clean up the contamination caused by the fire at Sandoz Chemicals in Basel in 1986. When the Schweizerhalle warehouse burned down, 9000 kg of pesticides (mainly phosphoric acid esters) as well as 13 kg of mercury (in the form of organic compounds) infiltrated the ground to a depth of up to 11 m. In order to clean up the site 43 000 ton of soil was

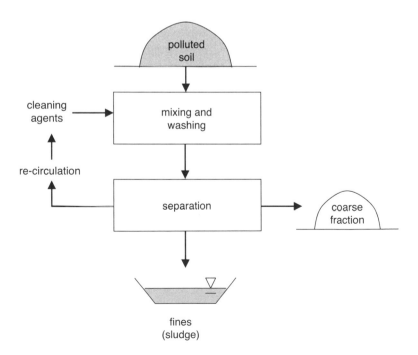

Figure 9.10 *Schematic material flow for soil washing*

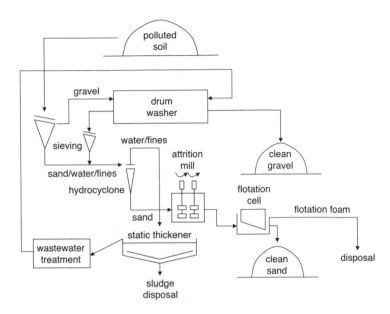

Figure 9.11 *Schematic material flows for a soil washing facility as employed for the Schweizerhalle accident in 1986*

excavated, of which 40 % had to be treated. Figure 9.11 describes a simplified soil washing facility. In the first step, the coarse fraction is separated by sieving. The separated gravel, being easy to treat, is washed with process water in a counter-current drum washer. From the remaining fraction, the fines are separated with a hydrocyclone. The isolated sand fraction passes through a series of attrition mills, where the grains are rubbed against each other to release the adsorbed contaminants. In floatation cells the contaminants are separated with the aid of surfactants. The fines, being the most contaminated fraction, are thickened in a static thickener and subsequently disposed of or treated with other technologies. The clean-up operation was completed within 26 months with a budget of 40 million euros i.e. 200 euro ton^{-1} of treated soil. It was the largest soil-washing endeavour in Switzerland.

Thermal treatment applies only to organic contaminants. In a typical thermal treatment operation the soil is de-volatilised i.e. dried and heated to a temperature of up to 400°C (*volatilisation*). In this phase, VOCs (volatile organic contaminants) that can be condensed are liberated and recovered to fuel the incineration process. In a second phase, non-volatile and problematic compounds like chlorinated hydrocarbons are incinerated at temperatures exceeding 1000°C (*incineration*), as depicted in Figure 9.12. Depending on the volume to be treated and the type of contaminant rotary kiln systems, fluid bed incinerators or infrared units may be employed. Since toxic flue gases are released during the incineration process, sophisticated pollution control systems have to be employed. The bottom

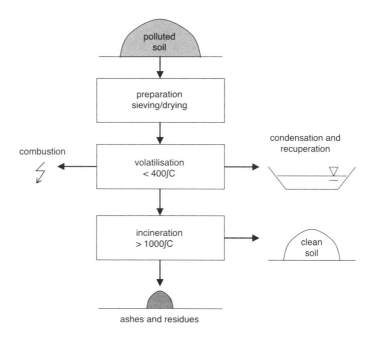

Figure 9.12 *Thermal treatment (volatilisation and incineration)*

ashes of the incinerator have to be analysed and possibly treated further. After incineration, the contaminated soil is clean but sterile from the biological point of view.

An emerging technology especially apt for SVOCs (semivolatile organic contaminants) and pesticides is *pyrolysis* i.e. the mineralisation of organic compounds in the absence of oxygen. Pyrolysis takes place at a temperature of about 430°C and breaks down organic components to combustible gases and solid residues. Due to the lower temperatures, the energy costs are reduced. The advantage of volatilisation, incineration and pyrolysis is that these technologies can be installed quickly on-site with mobile treatment units. Of disadvantage are, however, are the emissions to be treated, the high-energy input (especially for the incineration process) and the complete destruction of the soil structure.

Biological treatment processes apply to biodegradable contaminants. The biodegradation work is carried out by microbes, fungi and plants which transform the contaminants into daughter products (metabolites) and eventually mineralise them to form inorganic residues, carbon dioxide and water. Microorganisms capable of breaking down organic components are already present in natural soil. During a biological treatment procedure they are stimulated with nutrients and sometimes complemented with specialised microorganisms. If enough space is available, the contaminated excavation may be spread over an impermeable layer or on low permeability soil to farm and cultivate.

Landfarming operations, as depicted in Figure 9.13, aim at mineralising large volumes of contaminated excavation. Another option is soil heaping (Figure 9.14) i.e.

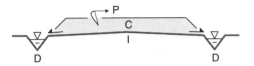

Figure 9.13 *Landfarming: C contaminated soil; D drainage; I impervious layer; P plough*

composting contaminated soil in stockpiles or windrows that may be several metres high. Infiltration of contaminated leachate into the ground is prevented by sealing the surface before heaping, with geomembranes for example. Non-toxic organic matter may be added to fuel the degradation process and maintain an elevated process temperature. Bulking agents are mixed into the soil to facilitate the aeration necessary for an accelerated biodegradation. The collected process water may be enriched with nutrients and recycled to further speed-up the operation. Biodegradation eliminates organic contaminants without much additional energy input, thus making this remediation option highly economic. Furthermore, the soil is not completely destroyed as it is in soil washing or thermal treatment. However, sufficient space has to be provided to allow landfarming or soil heaping. In addition, biodegradation processes may take a long time, sometimes years. Attention has to be paid to toxic metabolites that could be more toxic than the original contaminants. Nevertheless, biodegradation appears to be a smart option since the aspect of healing with natural agents is introduced. This aspect becomes even more important when biodegrading organic contaminants *in situ*, discussed below.

Isolation and stabilisation

Excavation of contaminated soil is not always feasible. The contamination source might be located too deep in the ground or underneath existing buildings. The contaminants may have already spread too far or migrated into the bedrock. Massive foundations left

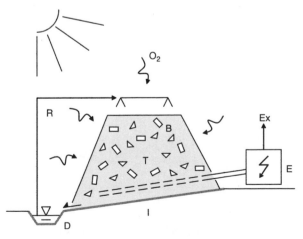

Figure 9.14 *Soil heaping: B bulking agents; D drainage; E energy; Ex exhaust; I impervious layer; R re-circulation; T thermal source*

in the ground may obstruct the excavation work. A shallow groundwater table may complicate the excavation. The risk that pollutants mobilise during the excavation process may be too high. The safety precautions to avoid a direct contact with the contaminated material may be too expensive, since they may include protection gear for workers, air filters for excavators, decontamination installations for equipment and vehicles, and fencing-in of the working zone. Dumping or after-treatment of polluted excavation may call for a budget much too high to realise the remediation project.

Isolation measures constitute an alternative approach to excavation. They aim at introducing barriers to confine the contamination source and thus trap and immobilise the pollutants. Vertical cut-off walls like sheet piles, slurry walls, soil mixing walls, grout curtains etc. are 'keyed' into a layer of low permeable soil to effectively confine the hotspot, as depicted in Figure 9.15. In order to avoid any leakage via the cut-off walls, the groundwater table within the confined area has to be kept lower than outside. This creates a hydraulic gradient towards the contamination source. Pumping from within the sealed-off area thus flushes the hotspot. The extracted water has to be analysed and possibly treated, creating long-term monitoring and maintenance costs. On the other hand, a fundamental advantage of this approach is that the contamination is not touched and remains isolated or trapped in the ground. Cut-off walls are quickly introduced at reasonable costs at most soil conditions. The confinement is almost complete and permanent, provided the low permeable stratum keying the wall proves

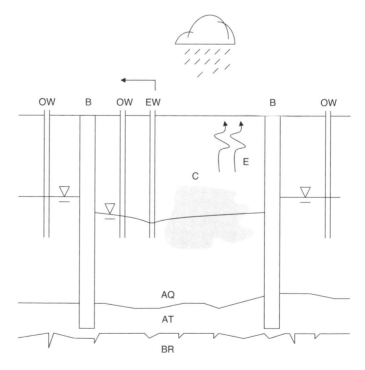

Figure 9.15 *Vertical barriers 'keyed' into an aquitard (low permeable layer): AQ aquifer; AT aquitard; BR bedrock; C contamination; E emission; OW observation well; B bentonite wall; EW extraction well*

continuous. Of disadvantage again is the fact that cut-off walls obstruct the natural groundwater flow, prompting problems in the vicinity such as damage to vegetation and structural damage to buildings due to rising and falling groundwater levels upstream and downstream of the sealed-off terrain. Gas emission from the contamination source may cause health problems and a complicated capping system may have to be installed. In fact, groundwaters monitoring wells and soil gas observation wells have to be installed inside and outside the keyed-off site.

Stabilisation of the contamination hotspot by injecting binding agents (Figure 9.16) offers another *in situ* alternative. This technique is also referred to as grouting. The injection material fills up the pores between the soil particles and stabilises them, thus rendering the contamination source impermeable and trapping the pollutants. A variety of injection agents is available including cement, pozzolanics and silicate gels. The choice of injection material depends on the type of soil and the contaminant to be immobilised. As a general rule, cements are injected in gravel and coarse sand, silicate gels in sands, and special chemical products in fine-grained sands and coarse silts. Injections may also be carried out to seal fractures in bedrock that possibly act as migration pathways for contaminants. Injecting a hotspot is a cheap and rapid remediation measure that consumes only a little energy and resources. It is carried out from the surface without disturbing ongoing activities at the site. Any contact with the contaminated soil is avoided. Hotspots below existing buildings can also be treated, by inclining the injection borings. However, as in the case of isolation, injecting a hotspot disturbs the natural soil conditions and impedes the groundwater flow. During the injection process contaminants may be mobilised. Complete knowledge has not yet been acquired and the question remains if over time the immobilisation capacity of the binding agent fades. Consequently, aftercare is necessary, which includes monitoring of the efficiency of the measure.

Fine silts and clays cannot be injected since the soil pores are too small. Hence, for fine and cohesive soils, other methods of introducing binding agents such as deep soil mixing are applied. A special technique to immobilise a hotspot in fine-grained soils is vitrification. An electric current is introduced by means of electrodes, which

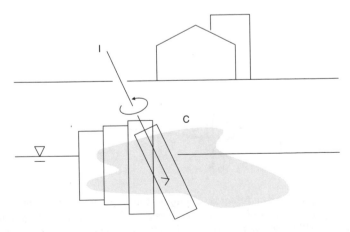

Figure 9.16 *Stabilising the contamination source with injections: C contamination; I injection*

vitrifies the soil at temperatures of up to 2000°C. While the soil vitrifies it decreases in volume, prompting the surface to subside. The energy required is much too high to allow this immobilisation technique to be commonly used. It is only applied in certain cases, for example when the soil is locally contaminated with radionuclides.

Hydraulic measures

Hydraulic measures apply to contaminants which spread with the groundwater. In order to eliminate these contaminants, groundwater may be extracted and subsequently treated, a technique that is referred to as 'pump-and-treat' (Figure 9.17a). For every pump-and-treat measure the hydrogeological context has to be studied and a model has to be established to decide how many wells are necessary, where they should be located and what capacity they should have.

Pump-and-treat is relevant mainly for contaminants that dissolve in the groundwater (aqueous phase liquids). But also non-aqueous phase liquids may be extracted. LNAPL (light non-aqueous phase liquids, Chapter 6) float on top of the groundwater table and thus accumulate in the depression cones of remediation wells where they can easily be drawn off with a second pump (Figure 9.17b). However, DNAPL (dense non-aqueous phase liquids, Chapter 6) sink to the base of the aquifer where they collect in depressions that are, in most cases, difficult to localise. The properties of LNAPLs and DNAPLs are listed in Tables 9.1 and 9.2 respectively.

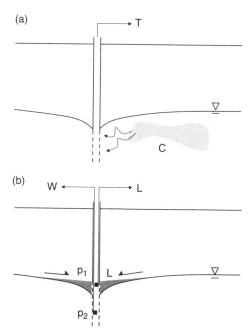

Figure 9.17 *A simple pump-and-treat system (a) with an extraction well: C contamination; T treatment and (b) the creation of a depression cone to extract LNAPLs: L light non-aqueous phase liquids (LNAPL); p1 pump extracting LNAPL; p2 pump extracting water to create depression cone; W water*

Table 9.1 *Properties of Light Non-Aqueous Phase Liquids (LNAPLs) and their remediation potential (Höhener 2001)*

compound	formula	density (20°C)	aerobic biodeg. potential	anaerobic. biodeg. potential	[b]aqueous solubility $c_{w\,sat}$ (mg l^{-1})	[b]vapour pressure $-logP_{vc}^o$ (atm)	[b]octanol-water partition coeff. log K_{ow}	[a]detection limit with nose (mg l^{-1})	conc. value OSC[d] (mg l^{-1})
Hydrocarbons									
Benzene	C_6H_6	0.879	+++	(-)	1790	0.90	2.13	0.17	0.01
Toluene	C_7H_8	0.867	+++	++	518	1.42	2.69	0.5–1.0	7.00
Ethylbenzene	C_8H_{10}	0.867	+++	+	168	1.90	3.15		3.00
p-Xylene	C_8H_{10}	0.860	+++	+	180	1.93	3.18	0.53	10.00
n-Hexane	C_6H_{14}	0.626	+++	+	12.70	0.69	4.11		2.00
n-Decane	$C_{10}H_{22}$	0.730	++	+	0.04	2.76			2.00
n-Hexadecane	$C_{16}H_{34}$	0.775	++	+	0.003	5.73			2.00
Gasoline (> 100 Compounds)	C_4–C_{10}	0.72–0.77	+++	(-)	100–240[c]	<3	2–4		
Kerosene (> 100 Compounds)	C_9–C_{13}	0.77–0.83	+++	(-)	15–25[c]	1–6	2–6		
Diesel Fuel (>100 Compounds)	C_{10}–C_{20}	0.82–0.86	+++	(-)	2–3[c]	3–7	3–7	<0.5	
Crude Oil (>1000 Compounds)	C_4–C_{40}	0.7–0.9	++	(-)	10–75[c]	1–9	2–8	<0.5	
Halo-organic compounds									
Chloromethane (Methylchloride)	CH_3Cl	0.92	+	++	5300	-0.76	0.91		
Chloroethene (Vinylchloride)	C_2H_3Cl	0.91	+	+	2790	-0.59	0.6		0.0001

[a] Rippen 2002
[b] At 25°C from Eastcott *et al.* 1988 and Montgomery 1996
[c] At 22°C from Shiu *et al.* 1990
[d] Ordonnance sur les sites contaminées 1998

Table 9.2 *Properties of Dense Non-Aqueous Phase Liquids (DNAPLs) and their remediation potential (Höhener 2001)*

compound	formula	density (20°C)	aerobic. biodeg. potential	anaerobic. biodeg. potential	[b]aqueous solubility $c_{w\,sat}$ (mg l^{-1})	[b]vapour pressure $-logP_{vc}^{\,o}$ (atm)	[b]octanol-water partition coeff. log K_{ow}	[a]detection limit with nose (mg l^{-1})	conc. value OSC[e] (mg l^{-1})
Haloorganic compounds									
Chlorobenzene	C_6H_5Cl	1.107	+		500	1.80	2.92	0.05	0.7
1,2-Dichloropropane	$C_3H_6Cl_2$	1.156	+		2700	1.18	2.28		0.005
trans-1,2-Dichloroethene	$C_2H_2Cl_2$	1.257	-(+)[d]	++	6300	0.35	2.09	0.0043	0.05
cis-1,2-Dichloroethene	$C_2H_2Cl_2$	1.284	-(+)[d]	+	3500				0.05
Dichloromethane (Methylenechloride)	CH_2Cl_2	1.329	-(+)[d]	+	19500	0.23	1.15		0.02
1,1,1-Trichloroethane	$C_2H_3Cl_3$	1.339	-	+++	1135	0.78	2.48		2
1,1,2-Trichloroethane	$C_2H_3Cl_3$	1.441	-		4500			50	
Trichloroethene	C_2HCl_3	1.465	-(+)[d]	++	1200	1.01	2.42	10	0.07
Trichloromethane (Chloroform)	$CHCl_3$	1.483	-	++	7700	0.59	1.93	0.1–20	0.04
Tetrachlormethane (Carbontetrachloride)	CCl_4	1.594	-	+++	970	0.82	2.73		0.002
1,1,2,2-Tetrachloroethane	$C_2H_2Cl_4$	1.596	-	++	3050	2.06	2.39	5	0.001
Tetrachloroethene	C_2Cl_4	1.623	-	+++	150	1.60	2.88	0,3–5	0.04
1,2-Dibromoethane	$C_2H_4Br_2$	2.172			1720	2.57	1.76		5E–05

Table 9.2 (continued)

Polyaromatic hydrocarbons (PAHs)

Naphthalene	$C_{10}H_8$	1.150	++	+	30.60	3.98	3.36	0.021	1.00
Anthracene	$C_{14}H_{10}$	1.240	+	-	0.73	8.10	4.54		10.00
Phenanthrene	$C_{14}H_{10}$	1.030	+	-	1.18	6.79	4.57	1	
Biphenyl	$C_{12}H_{10}$	1.180	+	-	7.00	5.00	4.09		
Coal tar	C_{10}–C_{24}	1.180	+	-	10-25[c]	4-9	3-9	<1	

[a] Rippen 2002
[b] At 25°C from Schwarzenbach *et al.* 2002 and Montgomery 1996
[c] Lee *et al.* 1992
[d] Degradable in presence of co-substrate (methane, butane, phenol, toluene)
[e] Ordonnance sur les sites contaminées 1998

The advantage of pump-and-treat is that no direct contact with contaminated soil becomes necessary. The contaminated groundwater may either be pumped to a wastewater treatment facility off-site or be treated on-site with mobile treatment installations. Remediation wells are readily installed at low costs even on urbanised terrain. The technology has been thoroughly tested and is generally accepted by both administration and public. Flushing circuits may be created by pumping contaminated water, treating it and re-injecting it upstream of the contamination source. Surfactants may be added to mobilise contaminants with a low solubility. It should be acknowledged, however, that the method is only applicable if the soil is sufficiently permeable and the pollutants use the groundwater as a migration pathway. Depression cones around the remediation wells may inflict upon the vegetation and trigger settlement of surface structures. It may be necessary to pump-and-treat for a long time until the contamination levels fall below desired threshold values. For example, at the Pintsch terrain in Hanau (Germany), dissolved contaminants were extracted with remediation wells, as depicted in Figure 9.18. The floating oil phase was pumped from depression cones around the wells. Biodegradation of organic contaminants within the unsaturated zone was stimulated with nutrients that were infiltrated from the surface (after Ripper 1992).

An alternative to pump-and-treat measures are reactive walls. They are installed in the affected aquifer downstream of the contamination source as passive *in situ* decontamination elements. Reactive walls are trenches that are filled with aggregates that react with the contaminated groundwater while it passes through them. Zero-valent iron walls are filled with sand and zero-valent iron that reacts with certain organic pollutants. Trenches filled with limestone and strongly alkaline material such as soda ash neutralise, for example, acid mine drainage. Adsorbing walls immobilise hydrophobic contaminants by adsorbing them to solid matter such as active carbon.

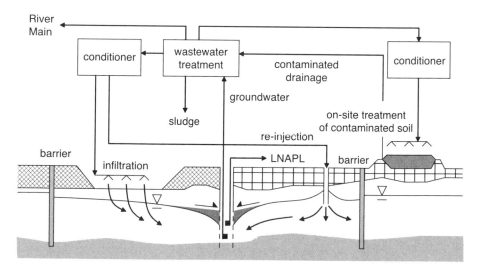

Figure 9.18 *Extraction of dissolved contaminants via remediation wells (after Ripper 1992)*

In bioreactive walls organic contaminants are degraded and mineralised by microbes. Aerating the reactive wall and adding nutrients enhances their performance.

There are two principle approaches to be distinguished: full reactive walls and funnel-and-gate systems. In the former, the trench is filled with reactive material at full lengths whereas in the latter case, low-permeable barriers are introduced which guide contaminated groundwater like a funnel to the permeable reactor or gate where the contaminants are eliminated (Starr and Cherry 1994). For low-permeable guiding walls, a permeability contrast of 1×10^{-3} to 1×10^{-4} has proven sufficient to achieve the funnelling effect. A good example of a funnel-and-gate system is the remediation of an aquifer contaminated with chlorinated hydrocarbons in Tübingen, Germany (Figure 9.19). The geological boundary conditions were favourable for a funnel-and-gate system, in that a 3-m thick loamy top layer overlies 6 m of sand and gravel, with a permeability of about 10^{-3} m s^{-1}. This stratum was contaminated with TCE and DCE. From a depth of > 9 m, low permeability loamy marls prevailed.

Both the reactive elements and the guiding wall have to be keyed into a low-permeable layer at the base of the aquifer to avoid contaminants passing below the installation. Since neither the contamination source nor the contaminated groundwater has to be extracted, further waste and wastewater treatment is unnecessary. Reactive walls are readily introduced at comparably low costs. Relatively few resources are consumed and only small amounts of waste are generated. Also of advantage is the fact that the natural groundwater flow is hardly affected. On the other hand, decontamination may take a long time during which the trench filling possibly has to be replaced to effectively degrade, neutralise or adsorb the contaminated groundwater.

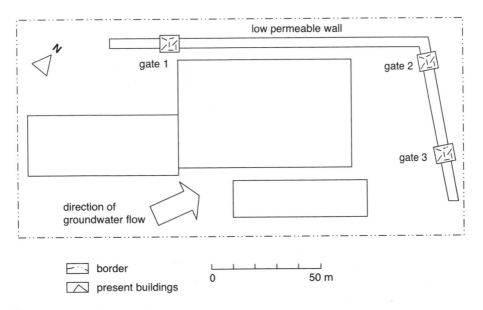

Figure 9.19 *Zero-valency iron reactors used to degrade chlorinated hydrocarbons (TCE, DCE) dissolved in the groundwater in Tübingen, Germany (after Schad and Teutsch 1999)*

Bioremediation

As mentioned already, organic contaminants degrade naturally due to the activity of microorganisms. This process can be enhanced by stimulating the activity of indigenous and inoculated micoorganisms with nutrients and electron acceptors (e.g. O_2, H_2O_2, NO^{3-}). Just as in the case of pump-and-treat, contaminated groundwater is extracted downstream of the contamination source and conditioned on site. In this case, however, the aim is to accelerate biodegradation. The conditioned, nutrient-enriched water is re-injected upstream of the contamination source, thus creating a closed circuit, a technique also referred to as the Raymond process (Figure 9.20), after R L Raymond and co-workers who first licensed this remediation strategy.

In situ bioremediation can be considered a flexible and inexpensive tool to degrade organic contaminants dissolved in the groundwater. Hardly any waste is produced as the contaminants are already mineralised by microorganisms. The advantages mentioned for the pump-and-treat approach also hold for *in situ* bioremediation. The utilisation of microorganisms renders *in situ* bioremediation even more efficient than regular pump-and-treat measures and reduces the time needed to effectively deplete the contamination.

Another method of stimulating biodegradation is the injection of air or pure oxygen into a contaminated aquifer. At the same time, nutrients may be infiltrated to further animate the degradation process. This pneumatic technique is called biosparging (Figure 9.21) and is especially suitable for the saturated zone. In order to extract the contaminated soil gas above the groundwater table, soil vapour extraction (SVP), also called soil venting when air is introduced instead of being extracted, may be applied (Figure 9.21). This technique is also suitable to volatilise organic

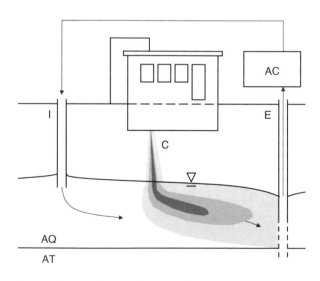

Figure 9.20 *In situ bioremediation (Raymond process): AC aeration and conditioning including the addition of nutrients; AQ aquifer; AT aquitard; C contamination; E extraction well; I infiltration well (re-injection)*

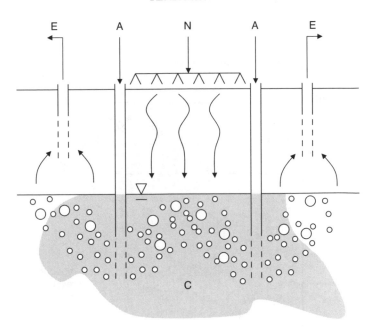

Figure 9.21 *Biosparging and soil vapour extraction: A air; C contamination; E exhaust (vacuum pump); N nutrients*

contaminants within the vadose zone i.e. above the groundwater table. Another technique to clean up the vadose zone is bioventing i.e. ventilating air or oxygen to stimulate the activity of indigenous bacteria that biodegrade the contaminants. Care has to be taken to avoid the development of preferred air channels that may develop after a certain operation time. To control this problem, only moderate air pressures are recommended. Biosparging and bioventing are usually applied to eliminate VOCs. When LNAPLs are extracted in a pump-and-treat approach as described earlier, soil vapour extraction and bioventing may also be applied to clean up the zone within the depression cone (which becomes contaminated when the cone forms). Note that the choice of appropriate bioremediation technique is dependent upon the physical characteristics of the contaminant to be eliminated; bioventing and soil vapour extraction are feasible only within a defined range of vapour pressures (Figure 9.22).

Natural attenuation

A contamination may be tolerated if neither the ecosystem nor other resources are threatened. In such cases, one may resort to simply observing the migration and intensity of the contamination. Since the contaminant naturally dilutes and possibly decays or biodegrades, the concentration of the contamination may fall to acceptable levels. It may therefore be sufficient to install observation wells and to probe them regularly to monitor the natural attenuation process, as depicted in Figure 9.23.

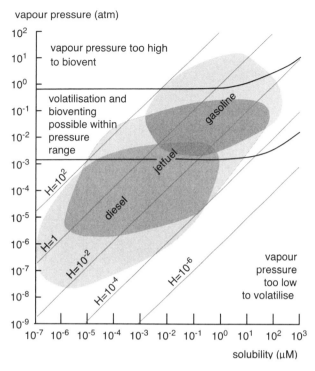

Figure 9.22 *The choice of appropriate bioremediation technique is dependent upon the physical characteristics of the contaminant to be eliminated (H, Henry's constant = 10^{-6} atm m^3 mol^{-1}) (after Höhener 2001)*

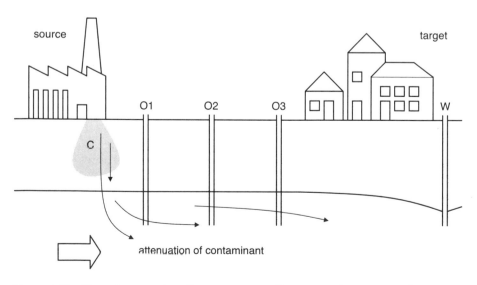

Figure 9.23 *Natural attenuation: C contamination; O1, O2, O3 observation wells; W water supply well (to be protected)*

If a contamination hotspot has been identified, for example a leaking tank, it may be extracted together with the soil in the direct vicinity, to eliminate the source. Subsequently, it may not be necessary to clean up the entire aquifer, which would call for a much larger budget.

The attenuation process usually takes a long time. Consequently, a long-term monitoring programme has to be implemented that is efficient enough to allow rapid reaction if the need arises. In general, a pre-warning buffer zone is established around the supposed contamination source where groundwater and soil gas sampling are carried out to screen the contamination status. This buffer zone must be large enough to allow enough time for intervention measures in case the contamination level unexpectedly increases. With a well-designed monitoring programme, the nature of the contamination will become clearer. An evaluation of the concentration of the contaminants over time indicates whether the pollution stems from a singular event, for example from a spill, or from a continuous input, for example from a leaking pipe. An apparent attenuation rate may be inferred that allows a prognosis on the required time until the contamination is no longer critical.

Intrinsic bioremediation of a heating oil spill in Studen, Switzerland (Bolliger *et al.* 1999)

In an urban area in Studen, Switzerland, a spill of more than 34 000 l of heating oil was discovered in an aquifer in 1993. The aquifer consists of unconsolidated glaciofluvial outwash deposits and is underlain at 20–25 m depth by an aquitard of molasse sediments. Hydraulic conductivities were determined with small single well pumping tests and range from 1.0×10^{-4} to 9.3×10^{-3} m s^{-1}. The groundwater table at the site is generally 2–4 m below the surface and slopes in a northeast direction with a gradient of about 0.15 m per 100 m. The temperature of the groundwater changes seasonally from a minimum of 9.5°C in March to a maximum of 11.5°C in September.

Small trenches were excavated down to the groundwater table to localise the contamination. Before refilling the trenches with clean soil, 11 wells were installed using prefabricated concrete pipes (0.6–0.8 m in diameter), which were screened along the lowest meter. Between 1993 and 1996, about 34 000 l of heating oil were recovered using floating pumps. After June 1996, no further action was undertaken. The source area covered about 90 × 30 m and remained stable. Chemical analyses of groundwater in the wells installed at the site revealed that anaerobic conditions prevailed, and that heating oil mineralisation was linked to the consumption of oxidants such as O_2, NO_3^-, and SO_4^{2-} as well as the production of reduced species such as Fe^{2+}, Mn^{2+}, H_2S and CH_4. Based on these results and the fact that there was no immediate threat to drinking water, the cantonal water protection authorities decided to limit engineered remediation efforts to the physical removal of the free heating oil phase. Instead of undertaking additional remediation efforts, natural attenuation processes were monitored. Analysis of dissolved inorganic carbon, the principal degradation product in the saturated zone, and its isotopic composition ($^{13}C/^{12}C$) allowed a better understanding of the mass balances associated with microbial heating oil degradation.

Figure 9.24 *Thlaspi caerulescens (left) and Iberis intermedia (right) hyper-accumulate metals (photographs courtesy of Catherine Keller)*

Phytoremediation

Plants extract nutrients and minerals from the soil in order to grow. We can take advantage of this faculty to extract shallow contaminants, a technique referred to as phytoremediation. For example: tobacco, Indian mustard and sunflowers accumulate metals. More than 300 plants are listed as accumulating nickel. Some 50 of about 175 species of the genus *Alyssum* are nickel hyper-accumulators. *Viola* species accumulate zinc, *Agrostis* species and *Minuartia verna* accumulate lead and *Thlaspi* species accumulate both (Figure 9.24a) whereas *Iberis intermedia* (Figure 9.24b) accumulates thalium. Other plants take up organic compounds such as pesticides, hydrocarbons (PAH, BTEX, etc.), polychlorinated biphenyls (PCB), chlorinated solvents (TCE, PCE) and propellants, explosives and pyrotechnics (PEP). Many plants may even take up radionuclides.

Phytoremediation is a complex process (see Figure 9.25) that has a number of different aspects (McCutcheon and Schnoor 2003):

- Phytoextraction: uptake of contaminants and the translocation of these contaminants either into harvestable aboveground plant tissues or into belowground root tissue.
- Phytotransformation: reduction and transformation of contaminants through the metabolism of the plant.
- Phytovolatilisation: release of extracted contaminants in volatile forms via the leaves of the plant.
- Phytostimulation (or plant-assisted bioremediation): animation of microbial and fungal degradation within the root zone.
- Phytostabilisation: holding of contaminated soil in place by vegetation and thereby immobilising toxic contaminants in soils.

Adding special fertilisers, mobilisers (chelating agents) and other additives enhances the performance of phytoremediation. Since there is a broad choice of plants suitable for phytoremediation, a number of discriminating criteria are applied. For example, plants should not accumulate noxious substance either in the roots (because they are too difficult to extract) or in the leaves (because they would then enter the food chain via the insects that feed on them). Rather they should accumulate in the

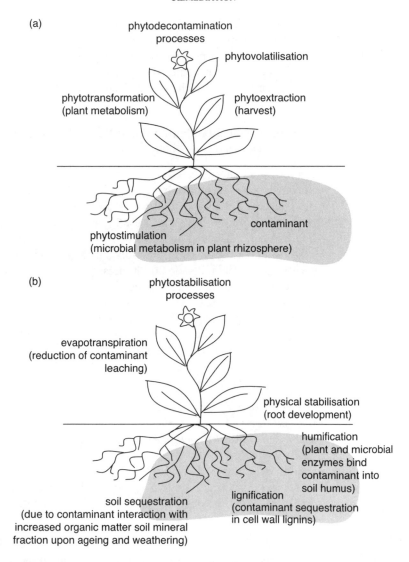

(a)

phytodecontamination
processes

phytovolatilisation

phytotransformation
(plant metabolism)

phytoextraction
(harvest)

contaminant

phytostimulation
(microbial metabolism in plant rhizosphere)

(b)

phytostabilisation
processes

evapotranspiration
(reduction of contaminant
leaching)

physical stabilisation
(root development)

humification
(plant and microbial
enzymes bind
contaminant into
soil humus)

lignification
(contaminant sequestration
in cell wall lignins)

soil sequestration
(due to contaminant interaction with
increased organic matter soil mineral
fraction upon ageing and weathering)

Figure 9.25 *Processes of phytoremediation (after Cunningham et al. 1996)*

stems to be readily harvested and further exploited, for example in energy production (incineration, biogas) or as construction material or other non-food products.

Phytoremediation has a number of advantages. It is a low-budget *in situ* approach utilising solar radiation as an energy source. It can be applied to a diffuse contamination affecting large areas. It is compatible with the natural environment and in line with the goal of restoring natural soil conditions. Many different contaminants can be addressed and even combinations of contaminants can be treated. Phytoremediation appears an acceptable method both for the administration in charge and the public. The harvested plants can be further exploited.

Of disadvantage, however, is the long treatment time that usually comprises several planting periods. The remediation depth is restricted to the root zone i.e. about 1 m, although deep-rooting trees such as poplars may reach greater depths. It may become necessary to apply additives that could, if not applied properly, mobilise contaminants. Phytoremediation is a new technique and not much experience has been made so far.

Phytoextraction of a heavy metal contaminated soil at Dornach, Switzerland (Kayser *et al.* 2000)

The surroundings of a brass smelter located in Dornach, Canton Solothurn, Switzerland, have been contaminated by atmospheric depositions of Cd, Cu and Zn over a period of about a century. The concentrations in the soil depend on the distance to the smelter and the wind direction. Land use in the contaminated zone includes agricultural and horticultural fields, house gardens, woods and meadows. The contamination is restricted to the upper part of the soil and reaches depths of up to 60 cm. The heterogeneity and present utilisation of the area makes the application of classical remediation techniques unsuitable. In some parts of the area, heavy metal concentrations have passed the clean-up values given by the Swiss ordinance relating to impacts on the soil and therefore require remediation. Within the framework of a multidisciplinary study, a field experiment was conducted 300 m NE of the brass smelter where total heavy metal concentrations are about 2.5 ppm, 650 ppm and 550 ppm of Cd, Zn and Cu respectively. The soil is a former *calcaric fluvisol* disturbed by road construction. The aim of the work was

- To test phytoextraction of Cd, Cu and Zn by hyperaccumulating plants (*Thlaspi caerulescens* and *Alyssum murale*), agricultural crop plants (*Nicotiana tabacum, Zea mays, Brassica juncea* and *Helianthus annuus*) and a woody plant (*Salix viminalis*).
- To improve phytoextraction efficiency using additives (assisted phytoextraction), namely nitrilotriacetate (NTA, a synthetic chelating agent) and sulphur S_8. This was tested because the heavy metal bioavailability was low due to a high soil pH.

The metal uptake was generally low for all plants without additives compared to results found in solution or pots experiments. NTA and sulphur increased the heavy metal solubility in soil by a factor of 21, 58 and 9 for Zn, Cd and Cu respectively. However, the accumulation in plants was only multiplied by a factor of 2 or 3. The highest concentrations were found in the hyperaccumulating plants (maximal 10 ppm Cd, 2500 ppm Zn and 40 ppm Cu) followed by *S. viminalis* for Cd and Zn (ca. 4 ppm and 300 ppm respectively) and *N. tabacum* for Cu (ca. 35 ppm). The total uptake was highest by *N. tabacum* (Cd) and *H. annuus* (Zn and Cu) because of their high biomass production. The authors concluded from the level of uptake that remediation would take more than 10 years assuming a 50 % removal of total metal from the soil. More promising results were obtained one year later with a total Cd removal 4.5 times higher for *T. caerulescens* (with 180 g ha^{-1}) than previously measured with *N. tabacum*.

Treatment trains

In most cases it may be advantageous to combine different approaches to optimise the remediation result. A typical example of a treatment train would be the excavation of the hotspot, the separation of the extracted soil into low contaminated and high contaminated portions and the subsequent treatment of problematic material (washing, thermal treatment, biological treatment, etc.). Another example is the previously mentioned combination of pump-and-treat with bioremediation measures plus the application of bioventing to clean up the area within the depression cones of the remediation wells.

An example of a treatment train for nonhalogenated VOCs, extracted while biosparging is applied to enhance biodegradation within the groundwater body, is depicted in Figure 9.26. The VOC-contaminated vapours are separated in a liquid/vapour separator. A granulated activated carbon adsorption system (GAC) is employed to separate the VOCs from the water, while the VOCs in the gas stream are destroyed by catalytic oxidation. Many combinations are possible to realise treatment trains for which resource consumption and waste production are minimised while remediation costs are kept low.

9.3.2 Acidified terrain

The amount of soil affected by acidification is difficult to assess. According to the GLASOD-survey less than 1 % of all human-induced soil degradation is caused by

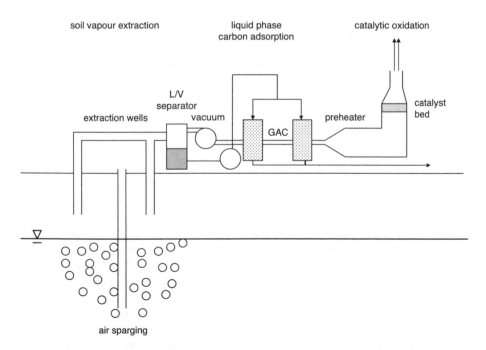

Figure 9.26 *Treatment train for nonhalogenated VOCs, extracted while biosparging is applied to enhance biodegradation within the groundwater body (after FRTR 2006)*

acidification (Oldeman *et al.* 1991). However, the ongoing deforestation, the excessive application of fertilisers and the still considerable acid rain-producing emissions suggest that the acidification problem may be more important than this study indicates. In addition, it has to be taken into consideration that, besides soils, lakes and groundwater resources have become acidic.

Acidification of soil may be avoided with measures of prevention. Once soil has turned acid, remediation measures become necessary.

9.3.2.1 Measures of prevention

The most straightforward approach to preventing soil acidification is a reduction of emissions producing acidification problems. This refers especially to acid rain producing emissions, which can be eliminated by means of modern filter technologies for automobiles and smoke stacks. In the last 30 years, considerable progress has been made in Europe and the United States. However, many countries including the former Soviet states and China as well as low-income countries all over the world continue to emit high quantities of emissions.

Halting deforestation is certainly a key issue when dealing with acidification. Once the vegetation cover has been weakened or destroyed, precipitation infiltrates the soil at an increased rate. As a consequence, nutrients are leached while being replaced by hydrogen, a process that renders soil acid.

On farmland, prevention focuses on the introduction of sustainable farming practices. With conservation agriculture, as promoted by the FAO, a permanent or semi-permanent organic soil cover is maintained and tillage is minimised. In addition, the amount of fertilisers applied is reduced to an amount that plants actually require. Further measures to reduce acidification are the relegation of slash-and-burn clearance, continuous and diversified crop rotation and the introduction of the rule that every hectare of deforestation should be compensated by at least one hectare of reforestation.

9.3.2.2 Measures of remediation

Remediation measures for acidified terrain focus on

- soil liming as an amelioration measure, to promote
- phytostabilisation of the surface.

Liming increases the pH of the soil and thus increases the availability of nutrients (Figure 9.27) while improving biochemical processes within the root zone. Liming also provides calcium and magnesium as nutrients for plants and increases the efficiency of nutrients such as nitrates and phosphates. These aspects of soil improvement eventually allow a vegetation cover to re-establish and biodiversity to redevelop.

As liming agents, ground limestone and dolomite are usually applied, which are also called *agricultural limestone* or *aglime*. For pelletised or granular lime, ground lime is compressed into pellets or granules to avoid dust problems. Granular lime does not, however, mix with the soil as easily as aglime does. Further alternatives are ground oyster shells, wood ash and lime sludge from water

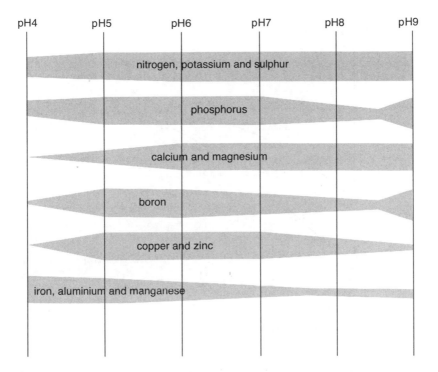

Figure 9.27 *Effects of change in soil pH on the relative availability of plant nutrients (after Spies and Harms 2005)*

softening plants. Fluid-bed ash, fly ash and stack dust are also applied, but should first be thoroughly analysed to ensure that they are not contaminated. Burned lime is very reactive and is therefore only applied in special cases as hydrated lime (or slaked lime).

The quality of the lime applied is dependent on its purity and the fineness of grind. Purity is expressed in calcium carbonate equivalent (CCE), with pure calcium carbonate having a purity of 100 %. Fineness is defined by the particle distribution, which is calculated from the mesh sieves that the ground lime passes through. Finely ground lime distributes more evenly and reacts more quickly with the soil. The purer the limestone and the finer it is ground, the higher is the neutralising capacity. The combination of chemical purity and particle size distribution gives rise to a variety of rating parameters including the effective neutralizing value (ENV), the relative neutralizing value (RNV), the effective neutralizing power (ENP), the total neutralizing power (TNP), the neutralizing value (NV), the neutralizing index (NI), the effective calcium carbonate (ECC) and the fineness factor (ALGLL 2004).

The rate of lime required depends upon the acidity of the soil, the type of the soil and its buffering capacity. In a lime requirement survey, the first step is to measure the acidity of the soil by mixing it with distilled water (Chapter 4). If the water-pH

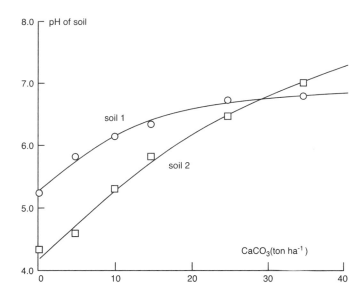

Figure 9.28 *Lime requirement curves*

drops below 7, the soil is identified as acidic. In a second test, the amount of lime required to sufficiently raise the pH is determined. Soil resists a change in pH due to its buffering capacity, which is a function of the clay content and the content of organic matter. Clayey soils have a much higher buffer capacity than sandy soils, and thus require more lime to change the pH value. The buffering capacity is expressed by the buffer pH (or lime index), which has to be determined in order to decide how much lime has to be added. This is usually done by mixing the soil with buffering solutions while measuring the change in pH. In lime requirement curves (depicted in Figure 9.28), the amount of lime required per hectare is plotted against the soil pH. Liming may have to be repeated several times until the functions of the soil are restored.

Lime is typically applied uniformly, for example with trucks equipped with spinner blades. If the pH is very low, lime may be mixed by tillage. A good time to carry out a liming campaign may be autumn, since the lime can dissolve over the winter to render the soil sufficiently alkaline before plants start re-greening and crops are seeded. However, the optimal liming time and procedure depends on many factors including soil types, soil use and weather conditions. Thus, the liming campaign has to be adapted to the local conditions and may include an extensive trial-and-error phase.

9.3.3 Salt-affected terrain

Worldwide, about 4 % of human-induced soil degradation is caused by salinization (Oldeman *et al.* 1991). Salinization may occur within the solum as *salin* (salt rich) horizon and at the surface as salt crust. Salty runoff may affect surface watercourses

The re-greening of Sudbury, Canada

The area around Sudbury has been heavily degraded due to metal mining (Chapter 6). The combination of clear cut, erosion and air pollution virtually wiped out all vegetation in the vicinity of mining and processing plants, leaving a 'moonscape' behind, in which even outcropping rock was altered chemically. However, some 25 years ago, the citizens of Sudbury decided to stop the destruction of their natural environment. In an unprecedented award-winning campaign the affected land was remediated utilising a remarkable spectrum of measures. The introduction of modern smokestacks, liming of land and water to reduce acidity and sophisticated re-vegetation strategies were the most important. At the Earth Summit in Rio de Janeiro in 1992, the re-greening program of Sudbury was recognised by the United Nations as an outstanding remediation exercise. Today, Sudbury looks completely different from three decades ago (Figure 9.29). Although mining has not ceased, the living quality has improved considerably. The methods and tools that were used to achieve this goal have been well documented by John M. Gunn (1995).

(a) (b)

Figure 9.29 *A site at the Copper Cliff tailings area (a) in 1973 (photograph courtesy of Tom Peters) and (b) in 1994 (photograph courtesy of Ellen Peale)*

and degrade aquatic biodiversity. It may render the water in rivers and lakes unsuitable for human consumption. Salt rich solutions may also reach the groundwater and spoil it.

9.3.3.1 Measures of prevention

Halting deforestation is an important part of challenging salinization problems. A vegetation cover reduces evaporation that would otherwise causes salt-rich solution to rise and spoil the upper active soil horizons. Furthermore, it regulates the depth of the groundwater table, which plays a major role in the salinization of soil.

On farmland, a promising measure of prevention is the introduction of sustainable farming practices. As in the case of acidification, conservation agriculture as promoted by the FAO is the key to avoiding salinization problems, since a permanent or semi-permanent organic soil cover is maintained and tillage is minimised. High salt index fertilisers are avoided and the amount of fertilisers applied is reduced to a minimum. In addition, irrigation of fields is optimised so that minimum evaporation occurs before water reaches the plants. Salt-tolerant crops such as rice, millet, palms and fodder (alfalfa) may be rotated to release salinization.

Since the groundwater table may wield salinization as it approaches the surface, all activities that may cause a rise of the groundwater table have to be reviewed critically. For example, in underground mining areas, subsidence should be minimised if it is known that salinization problems are likely. Pumping of groundwater close to the shoreline should be avoided, since this may trigger the intrusion of saltwater.

Further measures of prevention include the reduction of emissions which produce salinization problems. Effluents need to be desalinated before they are released into the environment. For example, salt-rich wastewater has to be treated before it is infiltrated into the ground and salt-bearing refuse, for example from salt mining, should be isolated from its surroundings to avoid saline runoff. De-icing salts should be used with care and replaced whenever possible by sand and cinders.

9.3.3.2   Measures of remediation

Remedial measures for saline soils focus on

- soil amelioration measures, to promote
- phytostabilisation of the surface and
- phytoextraction of salt.

Efforts to leach salt from the root zone (Keren 2000) have proven problematic since freshwater is usually scarce, especially in regions suffering from salinization. Besides this, other soluble components may be leached, prompting the pollution of surface waters and groundwater.

Ameliorating salt-affected soil may be achieved by adding organic matter and mulching. Spreading soluble calcium salt such as gypsum further improves the soil, with harmful salts being converted to less harmful salts. Measures of soil amelioration improve the capacity of the soil to develop a permanent vegetation cover and thus significantly reduce evaporation rates.

Besides direct soil treatment, a vegetation cover may be established by raising salt-tolerant plants. Shrubs and trees typical of the local environment and adapted to the given climate may be combined to minimise evaporation and optimise biodiversity. Phytoextraction appears to be a promising option to extract salt. Salt-accumulating plants cultivated to extract salt include *Suaeda fruticosa*, *Prosopis spicifera*, *Leptochloea fusca* and some of the rush family (Barrow 1994, Goudie 2000). However, care has to be taken when introducing specialised plants since they may invade and dominate the local fauna.

9.4   Physical degradation

As discussed in Chapter 7, physical degradation comprises compaction, surface sealing, overbuilding, subsidence and waterlogging. In this chapter, measures of preventing physical soil degradation and methods of remediation are introduced.

9.4.1 Compacted terrain

Worldwide, about 3.5 % of human-induced soil degradation is caused by compaction (Oldeman *et al.* 1991). Compaction may occur at the same penetration depths as the stresses, which have been introduced by surface loads compressing the soil profile. This may be a few decimetres as in the case of ploughing, or more than a metre for heavy vehicles. Geological loads such as Pleistocene glaciers or sediment layers that have been subsequently eroded reach much deeper into the ground. They pre-consolidate the sediment, an aspect that can be traced in the laboratory by means of oedometers (Chapter 4). This chapter focuses on methods for preventing human-induced compaction as well as remediation measures for terrain which has already been compacted by human activities.

9.4.1.1 Measures of prevention

The surface load and the soil type and water content determine the degree of compaction. With increasing load, the depth and intensity of compaction increases. Cohesive soils are more susceptible than coarse soils. The Proctor density defines maximum compaction at optimal soil moisture (Chapter 4). Loading the surface repeatedly increases compaction.

Given this, measures of prevention can be taken. Since surface loads induce compaction, prevention begins by introducing loads with caution. On farmland, heavy machinery should be employed with care and preference should be given to smaller, lighter farming machines where possible. Ploughing should be minimised for the sake of conservation agriculture. On grazing land, the density of livestock per hectare should be monitored and overgrazing should be avoided as the vegetation cover shields the ground and protects it from excessive wetting during rainfall. On forestland, felling should be carried out selectively and only patch cutting should ideally be permitted. Heavy lumber machinery should not be allowed to enter the forest. Lumber works should be halted at wet weather when soil moisture is high. On recreation land, skiing and cross-country riding should be restricted or banned.

In addition, prognoses of the compaction risk of soil should be carried out in order to zone the terrain into sensible and robust parts. Models dealing with soil compaction have been proposed (Horn 2004a, Horn and Feige 2003, Trautner *et al.* 2000) but research to clearly define compaction risks is still ongoing.

9.4.1.2 Measures of remediation

Remediating compacted ground is considered a difficult task that not always proves to be successful. Basically, two approaches are pursued (Horn 2004b)

- loosening the compacted layers mechanically
- stimulating natural processes that loosen the soil.

Mechanical loosening involves the application of tools that break the compaction. The simplest approach would be ploughing the compacted layers. This may, however, provoke an increase in density below the new plough level, thus creating a deeper and possibly more compact zone. In addition, the compacted soil is not loosened in *sensu strictu*, since it is just broken into smaller pieces that may remain compact, especially if the clay content is high and the water content optimal to form resistant and low-permeable chunks.

More sophisticated approaches have been introduced including slotting techniques i.e. introducing narrow slots of about a decimetre into the compacted layer. The slots are filled with the material previously loosed in such a way that the soil profile is sufficiently provided with air and water. With this measure, the environmental conditions for soil organisms are improved while the overall stability of the soil profile remains intact (Jayawardane and Steward 1995, Jayawardane *et al.* 1995). Plants can penetrate the compacted layer with their roots, accessing water and nutrients below the compacted zone.

The second option is allowing natural processes of loosening to take place. This includes the action of soil organisms to introduce bioturbation, the capacity of certain clay minerals to shrink and expand with changing moisture (especially montmorillonites) as well as the action of frost within the soil. Creating conditions at the surface to allow frost to penetrate, water to infiltrate and evaporate again to introduce shrinking cracks stimulates natural fragmentation processes but may, on the other hand, conflict with other measures of soil conservation. It has to be taken into consideration, though, that it may take a long time until natural processes lead to de-compaction. For example, the compaction left by moving Roman troops can still be traced at certain locations, although they date back two millennia. For example, close to Königsbrunn in Bavaria, Germany, the Via Claudia is still recognisable on aerial photos due to changes in growth patterns. Currently, rapid remediation methods have not been developed, which leads us to the conclusion that soil compaction has to be considered as irreversible damage to the soil.

9.4.2 Sealed and overbuilt terrain

In Europe, almost 10 % of the total land surface is overbuilt and sealed (EEA 2003). Sealing and overbuilding brings about a complete loss of almost all soil functions (Chapter 7). Because of this, efforts to mitigate the adverse effects caused by sealing and overbuilding have intensified during the last few decades. Furthermore, a variety of strategies of land recycling and restoration have been developed.

9.4.2.1 Measures of prevention

Prevention measures against soil sealing aim at reducing land consumption and recycling used terrain. In an effort to cut down urban sprawl, the German Government has decided to reduce land consumption from the present figure of around 100 ha day^{-1} to 30 ha day^{-1} by the year 2030. At the same time, derelict and partly sealed terrain will be improved ecologically, which includes active unsealing measures. In the United Kingdom, by 2001 more than 60 % of newly erected buildings were established on used land (Dosch 2003). Similar efforts to recycle land are reported from France and other countries of the European community. In order to stimulate land recycling efforts, some countries have implemented laws obliging investors who develop projects on natural or agricultural land to compensate this land consumption with re-naturalisation works on degraded and sealed sites.

246

In addition, the sealing rate on newly developed land has to be kept low. This can be done, for example, by increasing the number of floors of the planned buildings, thus rendering them more efficient with respect to land conservation. In addition, car parks, pavements and access roads can be designed to occupy less space. Courtyards can be built without sealing the ground. An upper limit of sealing grade can be fixed by the municipality in order to enforce measures against ground sealing. In the city centres, sealing grades will however remain high especially since urban planners have consented to reduce urban sprawl by rendering cities more compact while increasing their attractiveness (see Figure 9.30). This new trend towards a revival of city centres leads to an overall reduction of sealing grades in the peripheries and contributes towards the implementation of soil protection measures.

Efforts have been made to render ground covers permeable. The simplest approach is using gravel beds instead of asphalt or concrete slabs on terrain that is less frequented and which experiences loading to a lesser degree such as overflow parking, residential driveways and walking paths. Permeable covers can also be constructed with paving stones, bricks and interlocking stone plates laid in such a way that rainwater can infiltrate them. The paving is placed on a sand bed to facilitate infiltration that can attain up to 40 % of undisturbed soil (Losch and Cordsen 2004b). Another alternative is pre-cast concrete lattice pavers that are placed on the ground with vegetation establishing within the openings. Lattice pavers reach up to 60 % of the permeability of natural soil. As well as paving systems, geogrids may be applied to reinforce lawn spaces, gravel walkways, parking lots, etc. They are designed as high strength plastic grids made of recycled material and filled either with gravel or with soil mixtures to allow a grass cover to establish. Much research has been carried

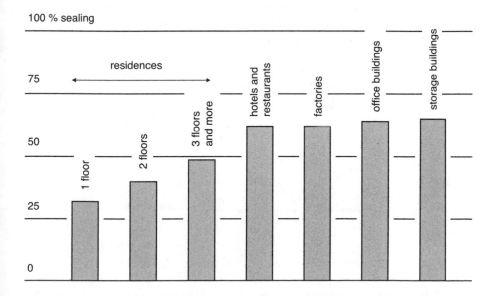

Figure 9.30 *Soil sealing rates in Germany 1995–2000 (Losch and Cordsen 2004a)*

out to develop permeable concrete and permeable asphalt pavements. Both have a smaller proportion of fines, making them pervious to rain and runoff.

Porous pavements have the advantage that stormwater runoff to be treated is greatly reduced since most of the runoff infiltrates directly into the pavement. A porous pavement typically consists of a porous surface underlain by a stone bed of uniformly graded and clean-washed course aggregate with a void space of at least 30 % (as depicted in Figure 9.31). The porous surface may consist of porous asphalt, porous concrete or pervious pavement units. A geotextile separates the aggregate from the underlying uncompacted soil. In northern climates less 'black ice' has been observed on porous pavements, and during rainy weather less water patches develop. Frost does not adversely affect the pavement since infiltrating water drains through the surface into the subsurface bed. In order to facilitate infiltration the underlying soil must not be compacted excessively. Pollution of porous pavements with sediments, fines and sediment-laden runoff must be avoided. Infiltrating contaminants like gasoline, oil and greases are readily biodegraded within the underlying gravel bed and the subsoil. Groundwater resources are thus recharged while contaminated runoff is reduced (PADEP 2005).

As an alternative to conventional architecture using flat footings to transfer the loads into the ground, pile foundations may be utilised as foundation elements since they disturb the soil only locally and may allow enough freeboard between the base of the building and the natural ground to maintain its ecological functions. The Greek historian Herodot reported on dwellings erected on wooden piles in the Lake Prasisas (today known as Lake Butkowo at the Struma in Macedonia) as early as the 5th Century BC. Since then, pile driving has become an acknowledged practice when buildings are erected on weak ground with low bearing capacity. Piles guide the loads to deeper, more bearable strata and may therefore carry much more than conventional flat footings, especially since steel and concrete piles have been introduced. If enough space is left between the surface and the base slab, the building is ventilated naturally

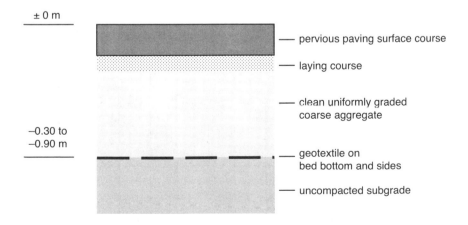

Figure 9.31 *Porous pavement, with porous surface underlain by stone bed of uniformly graded and clean-washed course aggregate (after PADEP 2005)*

and therefore kept dry and healthy. The elevation of the habitation protects the inhabitants from animals and mitigates the risk of diseases such as malaria, which makes piling a standard foundation method in humid climates. For example, the typical Japanese country house is founded on wooden piles with sufficient ventilation space between the floor of the house and the surface of the ground. In some cultures, public life happens below the piled dwelling. In Singapore, these public spaces were even furnished underneath newly built apartment blocks. Recovering the space underneath the habitation appears to be an appealing challenge to ecological architecture aiming at protecting the natural functions of soil.

9.4.2.2 Measures of remediation

Sealed and overbuilt land can be remediated with the purpose of

- recycling the terrain to restore its economical value
- re-naturalising the terrain to restore its ecological functions.

In the former case, degraded areas are reused to build new buildings and structures. Land recycling is a complex and usually expensive task that is, however, encouraged by all parties interested in soil protection. The remediation of industrially degraded land (e.g. Figure 9.32), also referred to as brownfields, directly reduces the need to build on ecologically intact terrain.

In the latter case, the site is upgraded ecologically. As mentioned before, re-greening may be an integral part of compensation measures that become necessary when ecologically intact sites or greenfields are overbuilt. In shrinking urban environments (Chapter 3), re-greening of brownfields appears attractive since the overall living quality of the city can be raised.

Figure 9.32 *Derelict terrain in the Ruhr District, Germany (photograph courtesy of Peter Drecker)*

Remediating sealed and overbuilt terrain involves:

- a comprehensive site investigation including a historical analysis to identify areas where the ground may be disturbed or contaminated
- an audit on the present state of the site, including an inventory of buildings, installations and underground structures
- an eco-compatible strategy to dismantle obsolete buildings
- strategies to either prepare suitable building ground for new buildings or to re-green the site.

On land that has been abandoned and that has been lying derelict for some time, surface features like roads and building fragments usually remain visible for a couple of decades. When buildings and surface structures are dismantled, the foundations and sometimes the entire substructures are left in the ground (for example, Figure 9.33). In certain cases massive underground structures may complicate redevelopment efforts (as shown in Figure 9.34). If the original building permissions are lost or incomplete, the localisation of fragments of former buildings and infrastructure may become a difficult task (Chapter 8).

Based on the site investigation work and an inventory of still existing structures, installations and fragments, ground obstacles may be grouped according to certain characteristics, including:

- size of the object
- type of the object
- potential to resist extraction
- age of the object.

Small objects are excavated easily, whereas large and massive foundations cannot be extracted using conventional methods. Historic brick foundations are readily removed whereas massive reinforced concrete foundations pose considerable difficulties. In certain cases, blasting may be the only way to destroy the obstacle. In urban areas, this may be both costly and risky. It has therefore proven advantageous to leave buried fragments in the ground and adjust the planning instead. Car parks or

Figure 9.33 *Foundation fragments and supply cables on a derelict site in Mediterranean Spain*

Figure 9.34 *Reinforced concrete foundation on industrial wasteland in Germany (photograph courtesy of Dietrich Merhoff)*

recreational parkland may be established on complicated ground. However, hollow structures like tanks, tunnels and canals have to be extracted or filled since they may cave in unexpectedly.

Another option is to re-utilise old foundations for new buildings. In this case, the new structure is erected on the old foundations. The bearing capacity of the foundation has to be investigated, with respect to both the bearing capacity of the ground and the strength of the foundation itself. Care should be taken when combining old and new foundations since new foundations settle under the structural loads whereas the ground underneath the old foundations is already compacted and therefore experiences much less settling. Differential settlement can be tolerated if the deformation of the ground does not cause cracks and instabilities.

Dismantling of structures is carried out following a comprehensive clearance strategy that includes (Genske 2003):

- a plan to meet safety regulations to protect workers
- a catalogue of derelict installations and machinery that can possibly be sold
- a description of decontamination tasks including drainage of tanks and pipes, evacuation of solid wastes, excavation of contaminated soil, removal of asbestos and other pollutants in buildings and installations, etc.
- a description of buildings and features to be protected due to their historical value and a strategy for avoiding damage to these buildings
- a description of buildings and installations that may be of use during and after the demolition work and a strategy for avoiding damage to these buildings
- concepts concerning the reuse, recycling and downcycling of demolished and excavated material
- a register of materials that cannot be re-utilised
- a survey on the stability of the buildings and facilities to be demolished, including structural details that may interfere with certain demolition techniques (for example, pre-stressed concrete elements)
- a demolition program including a detailed schedule indicating the sequence of breaking down structural components

- the parameters and circumstances that have led to the choice of certain demolition techniques, which may vary from simple crane-and-ball demolition, bulldozers and excavators to the application of explosives
- a programme to monitor the emissions caused by the demolition work, including aspects of soil and water contamination, air pollution (dusts and gases) and noise levels
- a plan to ensure a proper documentation of the clearance works, including a survey on the condition of the terrain before demolition
- a detailed financial plan, highlighting the benefits gained by recycling efforts
- a report on the eco-compatibility of all these measures i.e. a report on how the consumption of resources and the production of wastes are minimised and how environmental hazards during the operation can be avoided.

Recycling is a key issue of eco-compatibility and involves the separation of recyclables including iron, aluminium and copper scrap, cables and wires, windows and doors, wooden frames and supports, tiles and roofing paper. Paving material like pavers, stone plates and asphalt can be recovered and resold. Beds of compacted sand and gravel, which may have a reasonable market value, usually underlie a pavement. To this is added material that is downcycled to a lower economical value like crushed concrete aggregate from dismantled buildings and low-grade excavation. Recycling material that cannot be sold is preferably reused on-site to minimise waste streams leaving the terrain. Broken concrete and excavation may, for example, be used to remodel the site and to create bearable building ground. In addition, intact buildings and installations may be restored and put to new use. Restoring historic buildings is important since this maintains the heritage of the site and reminds the new user of how the site was utilised before it fell derelict.

The equipment used for the demolition work depends on the object to be dismantled. Removing a stone pavement is much easier than dismantling asphalt pavements and concrete slabs. Massive reinforced concrete foundations are almost impossible to remove. Care should be taken when using heavy machinery, since this causes compaction of the soil, thus degrading the terrain intended to be improved.

The site investigation work may reveal hints of a contamination beneath an overbuilt surface. In this case, dismantling and unsealing may conflict with efforts to protect resources like the groundwater. Certain parts of a derelict terrain may therefore remain sealed until the contamination has diminished or de-contamination measures have been decided upon.

Depending on the goal of the remediation work, either suitable building ground is established or re-greening work starts. In most land recycling projects, both become necessary. Since excavation and recycling usually produce considerable amounts of granular material, layers of bearable building ground can readily be provided. These layers are compacted to make the ground conditions more homogeneous and increase the bearing capacity. Ready-to-build land can therefore be promptly supplied, provided that underground obstacles have been removed, cavities have been filled and contamination hotspots have been eliminated.

It is much more difficult to restore environmental functions on land that was once sealed and overbuilt. When buildings were erected and pavements were laid, the upper horizons are usually removed because of their low bearing capacity. The ecologically active part of the soil profile is thus destroyed. The first step after removing the cover therefore aims at providing suitable conditions for plants to grow. Since the soil is usually compacted, measures of decompaction may become necessary, especially in cohesive soils. This can be combined with measures of soil amelioration, for example, by mixing organic material into the compacted layer while loosening it. Decompaction measures may prove difficult however, and a complete rehabilitation of soil functions may take a long time.

Substrates enriched with nutrients may be applied to stimulate plant growth. On derelict sites, obsolete cable canals, sewers and shafts may cause excessive dewatering of the site, disturbing the growth of the newly seeded plants. Therefore, potential drainage paths have to be mapped and possibly eliminated before starting the renaturation work. When new plants are seeded only local plants or those compatible with the local fauna

---

Un-sealing of a military base in Camp Reinsehlen, Germany (Blossey *et al.* 2005)

The 8-hectare terrain used to be an airstrip for the German Army during the Second World War. After the war, the site became a refugee camp until the British Army used it as a base and as training grounds. Tank shooting ranges were established. In 1994, the camp was handed over to the public and conversion work started. It was decided to re-naturate the terrain with the aim of creating the largest continuous site of *Corynephorus canescens*, a protected grass variety typical of dry, sandy soils. This involved extensive un-sealing work. In a first step, a thorough site investigation was carried out. Since no serious contamination was detected, the pavements were removed and stocked for reselling. 25 000 m^2 of stone pavement were sold at 2 euros per m^2. In addition, 37 062 m^2 of concrete driveway and 13 806 m^2 of asphalt pavement had to be removed. 3344 m^3 of refill soil had to be applied before the re-greening campaign could commence. Preference was given to the dry grass variety mentioned above, typical for the local environment. The costs of the un-sealing work amounted to 136 800 euros. By selling the pavement stones, some 50 000 euros could be recovered.

Demolition work in Oakalla, Canada (Weizäcker *et al.* 1997)

In 1991, the British Columbia Building Corporation advertised the demolition of a 24 × 46 m^2 concrete prison building by including the clause that each bidder had to propose two demolition strategies: a conventional and an alternative approach taking into account an optimised re-utilisation scheme for building components. The contract was granted to a bidder who proposed a demolition and recycling strategy with a budget 24 % cheaper than the conventional concept. The demolition contractor proposed to re-utilise three-quarters of the concrete blocks in a construction project of a local youth club, to recover and sell the wooden panels and girders, as well as windows, steel bars and other metal components. The composition of recycled demolition material is shown in Figure 9.35, which included gypsum walls, roof gravel that was sold to a landscaping architect and concrete rubble re-utilised as aggregate for a road project. Only 5 % of the demolition material had to be landfilled, in contrast to the conventional approach that would have landfilled some 90 %. The additional investment of one and a half working months for recovering and sorting was balanced by the income from selling the recycled goods.

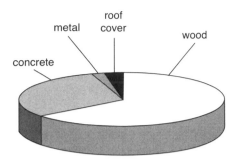

Figure 9.35 *Composition of recycled demolition material for the Oakalla*

should be used, and so re-greening of derelict land serves to restore the ecological function of the soil but also enhances the biological authenticity of the region.

9.4.3   Subsiding terrain

The worldwide extent of subsidence is not known. Subsidence of organic soils affects about 4.6 million hectares (Oldeman *et al.* 1991). To this, subsidence caused by mining and the abstraction of fluids and gases such as oil, water and natural gas has to be added. The cause of subsidence is always the removal of solids, liquids or gases, either by draining peat land, abstracting oil, water and gas or mining ore. This chapter focuses on prevention measures to avoid human-induced subsidence and remediation measures for terrain that may be subsiding or has already subsided.

9.4.3.1   Measures of prevention

Measures to prevent subsidence focus on compensating the voids created while the resource is extracted. This proves practicable in certain cases, however in most cases compensation measures are too complicated or expensive, or both.

In some cases, compensation measures even contradict the endeavour that causes subsidence. For instance, when peatland is drained to create farmland, subsidence is voluntarily accepted for the sake of cultivating ground. The only measure to avoid subsidence would be not draining the land i.e. accepting that the land cannot be prepared for agriculture. Searching for more suitable land, not subsiding and ecologically less vulnerable would be the only alternative.

When mining is the cause of subsidence, measures of prevention are possible though not always feasible. If solids such as ore are mined, backfilling or stowing the working may be an option. Stowing practices range from simple hand packing to mechanical backfilling with conveyers and high-speed throwing machines, hydraulic backfilling with water slurry and pneumatic backfilling with stowing dusts. Nevertheless, stowing does not completely eliminate subsidence. In fact, a reduction in subsidence of more than 50 % is rarely achieved. Moreover, stowing is expensive and may hamper the extraction work. In most cases, it is cheaper for the mining company to compensate the surface damage than to carry out an extensive stowing programme. Stowing is therefore only applied in rare cases. In this context, it has been proposed to dispose of waste material in idle workings. However, this practice has been rightly criticised as jeopardising groundwater resources.

Another option to reduce subsidence in mineral mining is partial extraction as carried out in room-and-pillar mining. Since the pillars support the roof of the mine, subsidence is prevented. However, when a room-and-pillar working is abandoned, it becomes almost impossible to monitor the performance of the pillars. Over the years, they may deform or even fail, especially if the mine becomes submerged and subjected to erosion.

Harmonic mining may moderate the surface effects of longwall panels. The basic idea is to superimpose two longwall panels in such a way that the strains at the fringe of the subsidence trough compensate each other. With this technique, the damage to engineering structures can be mitigated. The actual subsidence remains, however, as high as without harmonic mining.

If liquids such as oil are extracted, re-injecting fluids may counterbalance subsidence. This technique has frequently been applied, for example, at the Wilmington Oil fields in the US, where surveys have shown that even an uplift or rebound of the surface may be achieved with injections. Subsidence control wells may also be utilised to inject liquid wastes, a problematic practice since groundwater resources may be spoiled and gas emissions provoked. In addition, injecting fluids may activate geological faults and thus trigger earthquakes, as Bardwell (1970) reported from the Rocky Mountain Arsenal where noxious military compounds were mixed with water and injected into a major fault system. Recently, injecting water to produce geothermal energy triggered minor earthquakes in Basel, Switzerland (www.msnbc.msn.com/id/16128362/).

9.4.3.2 Measures of remediation

If measures of prevention have not been taken, subsidence will occur sooner or later. In this case, few options are left to remediate the terrain. Depending on the cause of subsidence and the depth of extraction, remediation measures may concentrate on

- halting subsidence with injections from the surface
- excavating the mined-out working.

If these measures were not feasible, a last resort would be to

- tolerate subsidence.

If shallow workings have been mined, filling of the voids may be subsequently accomplished from the surface by means of injections. This measure is just as expensive as stowing and is therefore only applied when certain areas at the surface have to be secured. The injection material usually consists of a mixture of cement, sand and water plus additives to render it stable against erosion. Even if the voids are

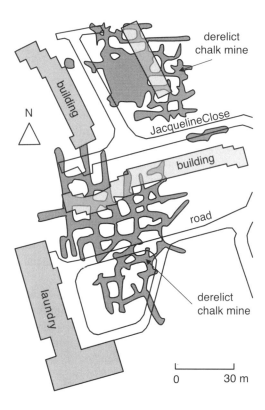

Figure 9.36 *The buildings of Jacqueline Close in St Edmunds Bury, England, eventually had to be dismantled to excavate the historic chalk mines that were the cause of sudden and irregular subsidence events*

completely filled, the contact zone between the injected material and the natural rock remains prone to erosion. Care has to be taken when several workings exist one above the other, since the loads due to the injection may lead to a collapse of the whole system.

The second more radical option is to excavate the terrain including the workings, refill it with bearable material and re-cultivate the surface. This has been done, for example, in the case of Jacqueline Close, a residential area in Suffolk, UK, developed over an abandoned chalk mine as depicted in Figure 9.36. Caving-in of the surface started about a hundred years after the mine was abandoned, and eventually forced the residents out of their newly built apartments. After the terrain was declared unsafe, it was stripped down to the bottom of the chalk mine at about 15 m below the surface and refilled with bearable soil. The site became a recreation area, while being monitored for ground movements (Bell *et al*. 2000).

The third option is accepting the fact that subsidence takes place. In this case, the subsiding area cannot be developed anymore. This allows natural spaces to become established, offering specific ecosystems that may enrich the local flora and fauna.

Subsidence lakes may develop, complementing the local opportunities for wildlife and biodiversity. If the subsidence is expected to be slow and regular, as for example in the case of deep longwall mining, the terrain may be open to the public and be utilised as recreation land. Alternatively, if the subsidence is expected to be irregular, as for example in the case of shallow room-and-pillar mining, the terrain should be closed to the public and designated a wildlife sanctuary.

Securing a casino in Valkenburg, Netherlands (Krapp and Vorwerk 1997)

In the course of site investigation for a casino building, voids were detected at 50–60 m below the surface. Historical documents showed that they are part of a medieval limestone mine. Many churches in the vicinity were erected with the material from this mine. The galleries were driven from the slopes of the mountain on top of which the casino was projected, following the 30 m thick limestone horizon of the upper Cretaceous (Maastricht). On top of this formation, some 25 m of Oligocene sands and a few metres of Quaternary loess-loams and gravels follow. In order to secure the future terrain of the casino, it was decided to drill into the openings and fill them with hydraulically binding mortar (Figure 9.37). The injection borings were spaced 8–9 m. Control borings were sunk after the injection work, in order to inspect the refill state and the contact between the injection material and the natural rock. Wherever necessary, a second injection round was ordered. After all the voids had been filled, the terrain was release for construction.

The development of Lake Lanstrop in Dortmund, Germany (Drecker *et al.* 1995)

In Dortmund-Lanstrop the Gneisenau Mine extracted five coal seams at depths between 300–500 m. The mine was decommissioned many years ago. However, subsidence has led to the development of a number of subsidence lakes, of which most were pumped empty or back-filled. The largest of these lakes, however, was left untouched. Lake Lanstrop developed between 1963 and 1967. It finally reached a width of 200 m and a length of 450 m. So far, the maximum depth of the lake is 9 m.

The formation of 'Lake Lanstrop' was accompanied by the subsidence of the old Friedrichshagener road (see Figure 9.38), which had to be diverted around the lake. Haus Wenge, an old moated castle situated adjacent to the lake, suffered structural damage but was restored as it is considered a monument of historic value. Sewers had to be reconstructed because subsidence reversed the gradient. Furthermore, agricultural was strongly affected by waterlogged spots.

However, the rapid formation of Lake Lanstrop also offered positive aspects. For example, it was of great interest to biologists who studied the return of plants and animals which had left the Ruhr District a long time ago. The new lake profoundly enhanced the living quality within the city. After a few years, Lake Lanstrop developed into a recreation park and a fishing centre, well frequented by the people of Dortmund.

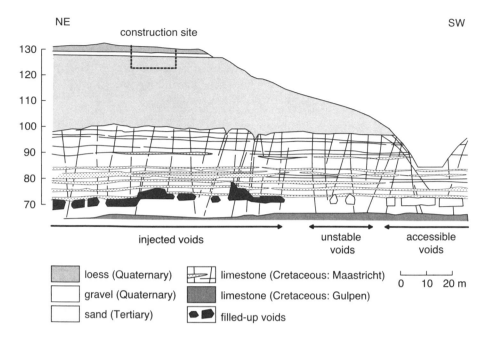

Figure 9.37 *The Valkenburg injection scheme*

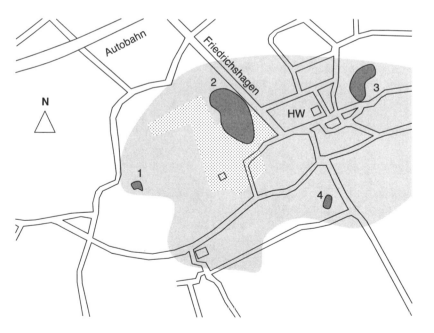

Figure 9.38 *The development of subsidence lakes in Dortmund-Lanstrop due to longwall mining: 2 Lanstrop-Lake; HW Haus Wenge*

9.4.4   Waterlogged terrain

About 11 million hectares of soil are degraded due to waterlogging (Oldeman *et al.* 1991). Waterlogging occurs when more water enters the upper soil horizons than they can drain. Human-induced waterlogging is mainly caused by soil compaction and subsidence. In addition, deforestation may trigger processes of soil solidification and laterization, which in turn may cause waterlogging (Chapter 6). Options to mitigate waterlogging can therefore be read of in the preceding sections.

10 Protection

In this final chapter, efforts and initiatives to protect the *resource land* are introduced. Special attention is given to local agenda processes, since they appear to provide the most efficient strategy of implementing the notion of sustainable land management. In addition, financing schemes are discussed and examples of good practice are presented.

10.1 From intention to routine

In 2004, the World Conservation Union IUCN stated that there are intrinsic factors and actual causes of soil degradation. Intrinsic factors are, for example, climate, terrain, vegetation, soil type and supported biodiversity. The actual causes of soil degradation are, on the other hand (Hannam and Boer 2004):

- biophysical impacts i.e. land use and land management, including deforestation, contamination, compaction, sealing, etc.
- socio-economic impacts i.e. human population increase, land tenure, marketing, etc.
- political forces i.e. pressure of incentives, ideology, legislation, etc.

In the preceding chapters, the types of degradation have been introduced as erosion, chemical degradation and physical degradation. Mechanisms have been explained and examples have been given. It has become clear that soil degradation is in most cases a slow process, not comparable with water pollution or air pollution that is immediately visible. For example, erosion may take place over a long period until fertility is lost. Chemical impacts are counterbalanced by the buffer capacities of the soil. The effects of physical degradation such as compaction accumulate over years, and consequences are only visible when it is too late.

In the preceding chapters it has become clear that measures of remediation of degraded terrain are in most cases complex and expensive. Fast solutions are rare and the rehabilitation of degraded wasteland can take decades. Measures of prevention have also been discussed, and it has become clear that we should aim for the sustainable management of land. The intention to take action about land degradation is certainly there; it has been proclaimed in countless speeches and declarations. The question is, however, how to implement sustainable land management practices in such as way that it becomes routine for everybody, accepted by both policy makers and the public, globally as well as locally.

10.2   Conventions, charters and protocols

The need for action to protect the resource land has been recognised by international institutions and organisations. The World Trade Organisation (WTO) has recognised the need for appropriate soil indicators for agricultural production and the Organisation of Economic Co-operation and Development (OECD) has began work on public policies for soil protection. International players including the Food and Agricultural Organisation (FAO) and other organisations of the United Nations, especially the United Nations Environment Programme (UNEP), have begun to formulate guidelines and recommendations for sustainable land management.

The Convention on the Protection of the World's Cultural and National Heritage was signed in 1972. In the same year, the *European Soil Charter* was presented, with the resolution (72) 19 as legal instrument. The *World Soil Charter* was adopted by the FAO in 1981 and by the UN one year later. Although not legally binding, it states

"...areas degraded by human activities shall be rehabilitated for purpose in accord with their natural potential and compatible with the well-being of affected populations."

UNEP proposed the *World Soils Policy* in 1982, based on the World Soil Charter. At the same time, the *World Charter of Nature* was presented, recognising ecosystems and biodiversity as fundamental resources.

In June 1992, the UN Conference on Environment and Development (which soon became known as the *Earth Summit*) was held in Rio de Janeiro. Representatives from nations all over the world discussed global environmental issues such as the decline of biodiversity, deforestation and climate change. The *Rio Declaration*, also known as the Earth Charter, was adopted, aiming to harmonise economical growth with environmental awareness. Inspired by the World Commission on Environment and Development, also known as the *Brundtland Commission* after its chief Gro Harlem Brundtland, a complex action plan referred to as *Agenda 21* was presented. The basic conclusion was to

"...ensure that [humanity] meets the needs of the present without compromising the ability of future generations to meet their own needs."

The notion of *sustainable development* was born, offering a realistic and robust concept to cope with the environmental challenges of the post-industrial era. Chapter 10 of the Rio Declaration focuses on an integral approach to planning and management of land resources:

"Land is a finite resource, while the natural resources it supports can vary over time and according to management conditions and uses. Expanding human requirements and economic activities are placing ever-increasing pressures on land resources, creating competition and conflicts and resulting in suboptimal use of both land and land resources. If, in the future, human requirements are to be met in a sustainable manner, it is now essential to resolve these conflicts and move towards more effective and efficient use of land and its natural resources."

Issues of soil protection are only addressed, however, in indirect ways with recommendations on sustainable land use and good agricultural practices. In addition, issues of biodiversity, desertification and socio-economic restrains are raised.

The work of the Brundtland Commission constitutes the basis upon which more specific conventions were launched including the *Montreal Protocol* on Substances that Deplete the Ozone Layer (1989), the *Basel Convention* on Trade in Hazardous Wastes (1989), the *Stockholm Convention* on Persistent Organic Pollutants POP (2001) and the *Rotterdam Convention* on the Prior Informed Consent Procedure for Certain Hazardous Chemicals and Pesticides in International Trade (2004). In 2005, the *Kyoto Protocol* to the United Nations Framework Convention on Climate Change came into force, which identifies soil as one of the major sinks of greenhouse gases.

In 2003, the *European Soil Charter* for the Protection and Sustainable Management of Soils was updated and revised. In addition, the European Union has been addressing the issue of soil protection in indirect ways with policies, perspectives and directives. Typical examples are the common policies for agricultural, water and waste as well as guidelines for environmental impact assessments and spatial development policies. Another example is the directive on the conservation of natural habitats and of wild fauna and flora.

Member states implement the directives of the EU in their own legislation. A typical example is the German Soil Protection Act (BBodSchG) that aims at conserving the resource soil by preventing degrading impacts and remediating degraded terrain. An interesting aspect of the German Environmental Protection Act (BNatSchG) is the obligation of compensating for greenfields that are built-over with the remediation of degraded terrain.

Re-greening of the pioneer barracks in Raila, Thuringia, Germany (Genske and Ruff 2006)

This complex was used by the youth organisation of the German Democratic Republic as a resort from 1964. In 1981, it was promoted to the 'Central Pioneer Resort Hermann Matern' and included several concrete panelled multi-floor buildings, a gymnasium, fabrication halls, 67 bungalows and an assembly ground. The site, located close to the village of Raila in southern Thuringia, was used at times by up to 1000 young pioneers.

After the reunification, the 16-ha site was no longer used but instead abandoned (see Figure 10.1). It was sold to a private investor but his plans to revitalise the terrain failed. The buildings began to dilapidate but the owner refused to carry out any restoration or demolition work. He was worried he would find contaminated sectors, which he would be required to clean up by law. Although the community encouraged him to re-green the terrain he declined. By then, the neighbouring city of Saalburg became interested in the site. As the construction of the new Autobahn A9 was underway, sites were being searched for to compensate for the destruction of natural land. According to German law, development projects on natural land must be compensated for by re-naturalising degraded land (§19 BNatSchG). Since the city council was in favour of the new Autobahn, it contacted the planning office (DEGES) offering them the derelict pioneer barracks. The DEGES accepted and the re-greening of the site began. The campaign involved three basic steps

- In the first step, the buildings were dismantled to ground level. Foundation plates were broken and perforated to re-establish infiltration.

- Thereafter, the site was modelled in earth to harmonise it with the surrounding natural land.
- In the third step, the land was re-greened. In some parts meadows were seeded, in others trees typical for the region were planted.

The financial requirements of the project were of the order 1.3 million euros, of which 0.8 million euros was spent on clearance work.

Figure 10.1 *The Raila Barracks before re-greening (DEGES)*

10.3   Networks and local agenda processes

In 1992, the year of the Earth Summit, Donella H. Meadows, Jorgen Randers and Dennis L. Meadows published a revision on their state-of-the-world scenario *The Limits of Growth*. In their new book they made a devastating prediction:

"In 1971, we concluded that the physical limits to human use of materials and energy were somewhat decades ahead. In 1991, when we looked again at the data, the computer model, and our own experience of the world, we realised that in spite of the world's improved technology, the greater awareness, the stronger environmental policies, many resource and pollution flows had gone beyond their sustainable limits. The conclusion came as a surprise to us…The human world is beyond its limits."

The Limits of Growth was recently updated once more, to underline the trends and tendencies that had been observed more than 30 years ago. Although the book has been criticised as being too pessimistic, certain conclusions such as those about resource land appear to be realistic. In addition, continuous efforts to undermine projects of environmental protection, such as the recent attempt to open the Arctic National Wildlife Refuge in Alaska for oil drilling, make it clear that there is not much political backing where profits are concerned.

Under the impression of a world of deteriorating resources, frustrated by the limited effects of conventions, charters and protocols, people began to build networks to take action on land protection on a local and a regional scale. For example, in 1996 the *European Soil Bureau* (ESB) was founded as a network of centres of excellence and national institutions. The goal of the ESB, stationed in Ispra (Italy), is to provide scientific information on land degradation for the EU and to develop the *European Soil Information System* (EUSIS).

In 1998, at the Congress of the International Union of Soil Scientists in Montpellier (France), the *Working Group on International Action of Sustainable Use of Soil* (IASUS) was formed. In 1999, the *European Soil Forum* (ESF) had its first meeting in Berlin (Germany). The goal of the ESF is to pool information in order to stimulate discussions and proposals of action for soil protection. Although a EU initiative, it also includes Switzerland and central and eastern European countries.

The *European Land and Soil Alliance* (ELSA) emerged from the *Climate Alliance* in 2000. ELSA is an association of cities, towns and rural districts with the aim of making an active contribution of sustainable land use. In October 2000, ELSA passed the *Manifesto for the Soil and Land Alliance of European Cities and Towns* in Bolzano (Italy). By accepting this manifesto the members commit themselves to a determined approach in terms of soil protection and spatial development. ELSA is in line with the idea of promoting *Local Agenda Processes*, which recognise that local communities are ideally positioned to take the lead in achieving ecologically sustainable development. In the same context, the *Tutzing Initiative* emerged in Tutzing (Germany).

As well as these networks, several concerted actions have been initiated. Of significance are the *Network for Industrially Contaminated Land in Europe* (NICOLE), the *Concerted Action on Risk Assessment for Contaminated Sites* (CARACAS), the *Contaminated Land Rehabilitation Network for Environmental Technologies in Europe* (CLARINET) and the *Concerted Action on Brownfield and Economic Regeneration Network* (CABERNET).

10.4   Structural funds

Agenda 21 aims at promoting sustainable management of land based on an integrated approach:

> "Land resources are used for a variety of purposes which interact and may compete with one another; therefore, it is desirable to plan and manage all uses in an integrated manner. Integration should take place at two levels, considering, on the one hand, all environmental, social and economic factors (including, for example, impacts of the various economic and social sectors on the environment and natural resources) and, on the other, all environmental and resource components together (i.e., air, water, biota, land, geological and natural resources). Integrated consideration facilitates appropriate choices and trade-offs, thus maximising sustainable productivity and use."

The underlying strategy is thus a combination of environmental, social and economic factors. In that respect, the Superfund-legislation of the United States is not a good example. In 1980, a federally controlled fund, the so-called *Superfund*, was installed under the *Comprehensive Environmental Response, Compensation and Liability Act* (CERCLA) to enable

- short-term removals [of contaminants] where action may be taken to address releases or threatening releases requiring prompt response
- long-term remedial response action to permanently and significantly reduce the danger associated with the releases or threats of hazardous substances that are serious, but not immediately life threatening.

The latter actions can be conducted only at sites listed on the *National Priorities List* (NPL) of the United States Environmental Protection Agency (EPA). According to Williams (1995), however, the NPL in practice has had the effect

"of causing the site, plus much of the adjacent area, to be abandoned for any use, and severely depressed property values."

In contrast, European funding schemes like the *European Regional Development Fund* (EFRE) aim to stimulate local economies while remediating degraded terrain and protecting virgin land. This involves parties at community level, such as possible investors who provide financial assistance since the EU commission only pays a part of the remediation costs. This in turn promotes the cooperation of public offices with private investors, a model referred to as *Public Private Partnership* (PPP). In addition, new and innovative projects in line with the idea of sustainable land use are promoted, an approach that encourages technological development (Genske 2003). Awareness is raised of how the site was used before it was abandoned and redevelopment plans were made. In the Garden Osterfeld in Dortmund, 'Time Walls' with the historic panorama engraved let the beholder merge the past with the present (Figure 10.2).

Figure 10.2 *Transparent 'Time Walls' with the historic panorama engraved lets the viewer merge the past with the present in the Garden Osterfeld in Dortmund, Germany (photograph courtesy of Peter Drecker)*

Practically, this means that the European strategy is stimulating an auto-empowerment of those concerned: the municipalities struggling to attract investors, the investors searching for cheap development sites, the ecologist fighting for the preservation of nature, the environmentalists trying to introduce new and clean technologies of resource utilisation and the unemployed searching for job opportunities. The projects of the *International Building Exhibition* (IBA) *Emscherpark* (1989-1999) typically followed this strategy. Local agenda processes were topped off with EFRE, national and regional funds, in which the leading role of the local community in achieving ecologically sustainable development goals was recognised. These goals were to be reached through integrating environmental, social and economic targets.

The re-vitalisation of the industrial site of Mont Cenis
(Genske 2003, Genske *et al.* 1999)

Mont-Cenis was originally a coalmine that was established in 1871 in the vicinity of Herne in the German Ruhr District. In 1893, the coal washing building was installed. Twelve years later the first coking facility was constructed. The mine prospered and became one of the biggest in the Ruhr District with 1750 company-owned housing units for its miners. For many generations, coal was extracted from longwall workings and refined on-site. In 1969, the 1300-m deep shaft of Mont-Cenis was the deepest of the region. The mine, which had a record production of 1×10^6 ton of coal in 1975, closed down just 3 years later.

The 26-ha terrain was abandoned; the negative attributes of the urban wasteland such as soil contamination, acid mine drainage, subsidence and massive underground structures obstructed any reuse of the site. In 1990, more than 120 years after the mine was founded, Mont-Cenis became one of the largest projects of the ambitious North Rhine-Westphalia program to remediate derelict industrial wasteland. The aim was to establish on the very same land new companies and enterprises, an important and urgent program since the unemployment rate in the Ruhr District exceeded 15% at that time. Because of the innovative character of the project, Mont-Cenis became part of the IBA Emscherpark. Furthermore, the project participated in the EXPO 2000. Aspects of this project were presented at the Biennial of Architecture in Venice and the World Climate Summit in Kyoto 1997.

In 1991, an international architectural competition on the re-utilisation of the site was organised by the IBA Emscherpark and the Land North Rhine-Westphalia NRW. The project included (EMC 1998):

- the construction of an academy for the Minister of the Interior of NRW
- public service buildings for the town of Herne including a multi-purpose meeting hall, civic administration buildings and a library
- additional shops and services for the existing shopping centre
- 250 housing units
- a recreation park,

as depicted in Figure 10.3.

In 1991, the architects Jourda & Perraudin (Paris) won the competition by proposing a tent-like structure made of glass and timber that would house the academy and many other facilities (see Figure 10.4). The utilisation of wood was proposed because of its resource efficiency, an aspect treated in more detail in the report for the Club of Rome of 1995 (Weizäcker *et al.* 1997). Only wood from the forests close to the construction site was used in order to reduce transport costs and to avoid exploiting foreign forests. In a research project supported by the European Union, Jourda & Perraudin Architects proved (together with their partners HHS Planer & Architekten, Kassel, Ove Arup & Partners, London and MTI France) that the tent structure creates a microclimate, which drastically reduces energy costs.   23 % of energy for air conditioning and heating is saved, resulting in a net reduction of 18 % CO_2 emissions. The microclimate created in the building reduces the energy consumption per m^2 to 32 kWh yr^{-1}.

To cover the structure, a roof-integrated solar power plant was installed, the largest in the world at that time. Some 10 000 m^2 of solar cells produce 0.75 million kWh yr^{-1} i.e. more than twice the energy needed to operate the centre. The solar panels function like clouds, shading the concrete structures inside the building so that they do not heat up during daytime but maintain a constant room temperature. A special battery facility was constructed to provide energy during periods of peak consumption.

As well as the utilisation of active and passive solar energy, mine gas escaping from neighbouring shafts is utilised. Every year, $120 \times 10^6 \, m^3$ of methane are emitted from decommissioned coalmines in the Ruhr District. This amount of mine gas is equivalent to some 100 000 tonnes of fuel oil. The general practice is to burn the gas without utilising it, resulting in emissions of about 8 Mton CO_2. For the first time in the Ruhr District, mine gas from an abandoned mine is being exploited to produce energy for local consumption (Figure 10.5). Mont-Cenis emits approximately $1 \times 10^6 \, m^3$ mine gas containing 80 % methane every year. The gas is converted into 9 million kWh yr^{-1} of electrical energy and 12 million kWh yr^{-1} of heat is directed to adjacent buildings. This practice results in a reduction of carbon emissions by 60 000 tonnes each year—a small but necessary contribution to combat the greenhouse effect. However, the emission of mine gas is not constant but fluctuates with atmospheric pressure. A 1.2 MW battery plant was therefore constructed to secure a constant energy supply. For periods of peak consumption, the installation was also connected to the municipal energy system. This connection may also be used to discharge surplus energy produced with mine gas.

Rainwater from the roofs of the buildings is collected and stored in ponds and cisterns. The stored water is utilised in the academy building for cleaning the solar panels, flushing the toilets, and watering the gardens both inside and outside the building. The ponds ensure the availability of constant supplies of water. The living quarters in the southeast of the oval vista are also equipped with rainwater storage facilities. Excess water is only infiltrated into the ground at places where the contamination of the soil is low enough so that the aquifer will not be polluted.

In the northern part of the project, the soil is contaminated. An excavation of contaminated soil was not considered since this would trigger an uncontrollable release of contaminants and require that expensive off-site treatment procedures be implemented. The storage of non-degradable wastes in landfills would also be required, which would occupy space and contribute further to land consumption. The alternative solution was to line contaminated sectors with membranes or clay liners, on top of which herb gardens were established. With this sealing measure, any further leaching of shallow pollutants into the groundwater would be avoided. Above the liner, a gravel and sand filter is installed to collect the precipitation.

The overall investment for this project was of order 110 million euros. The costs were shared between the Land North Rhine-Westphalia, local investors and the European Community. This mixed financing plan is characteristic of all rehabilitation projects subsidised by the European Community, calling for the active participation of the community and private investors.

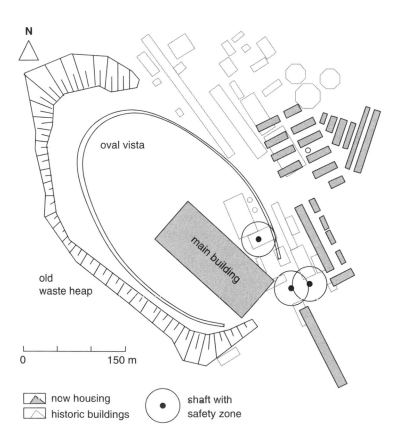

Figure 10.3 *Map of Mont Cenis (simplified)*

Figure 10.4   *Glass façade with solar roof and shaft cap (photograph courtesy of Thomas Lichtensteiger)*

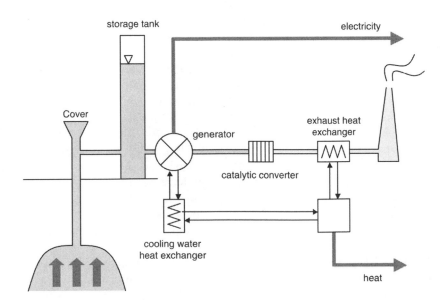

Figure 10.5   *Utilization of mine gas (after EMC 1998)*

10.5   Three principles

The first steps have been taken: the protection of land as a degrading but finite resource has become an issue of global concern. Many organisations and agencies have placed the topic at the top of their agenda. Networks have formed and concerted actions have been launched, both globally and locally, introducing policies and measures of sustainable land management. It is now important to encourage, harmonise and optimise these efforts.

As conventions, charters and protocols can only provide a general framework, the three areas involved in the challenges of land protection would appear to be:

- Visualising soil degradation processes to raise the awareness of governments, organisations and the public
- Promoting local agenda processes that combine land remediation with social and economic targets
- Creating tools of legislation and standards that support these processes in an efficient and transparent way.

Land inventories are the basis of land protection. Efforts to make degradation problems visible have already begun with the GLASOD Initiative, an attempt to globally assess the degree of soil degradation. The GLASOD Initiative has been maintained and used as an information system by UN organisations as well as many governments, agencies and institutions. Complementing these efforts, the *World Soils and Terrain Digital Database* (SoTer) is in the process of being composed. SoTer aims at developing national capabilities to provide geo-referenced soil and terrain information including soil types, erosion risks, soil carbon stocks, etc. The programme is carried out under the auspice of UNEP. The FAO in Rome, the International Soil Reference and Information Centre (ISRIC) in Wageningen (Netherlands) as well as the International Union of Soil Sciences (IUSS) are involved in the evaluation of the database and its implementation. In addition, the European Environmental Agency (EEA) is establishing the *European Environmental Information and Observation Network* (EIONET). 44 land cover categories have been established under the CORINE program to recognise the type of land cover and to identify changes in land use over time. Furthermore, indicators to derive environmental risk maps have been produced (EEA 1999). With advancing technologies, especially in the sector of satellite imaging, and decreasing costs of data acquisition and processing, land monitoring and land information systems have become a common tool of visualising land degradation problems.

Local agenda processes boost the implementation of sustainable land management at a grassroots level. Government and inter-governmental organisations are advised to support these processes. In addition to this, local agenda processes harmonise sustainable land management with social and economical aspects. They involve the people directly affected as well as local organisations, agencies and investors, thus directly promoting public-private partnership initiatives.

Creating tools of legislation and regulation is possibly the most difficult part, since some of these tools may end up obstructing local efforts by overly complicating procedures and routines. The World Conservation Union (IUCN), a network of governmental agencies and NGOs based in Gland (Switzerland), evaluates the feasibility of draft legislation for sustainable soils to aid nations concerned about their soil resources. Initiatives like this may ease the implementation of governmental directives on land protection.

Finally, the continuous effort of raising awareness is an integral part of sustainable land management. We also call upon the scientific community to continue making proposals to utilise land with respect for nature and for the benefit of us all. National and international organisations and agencies have already completed a great deal of important work. Their efforts need to be supported and strengthened, however, as there exists a powerful lobby which regard land as a resource to be exploited.

References

Adushkin V. V., Rodionov V. N., Turuntaev S. and Yudin A. E. (2000) Seismicity in oil fields. *Oilfield Review*, 12, 2–17.

AGB (2005) Soil mapping manual (Bodenkundliche Kartieranleitung). AG Boden, Schweizerbart Stuttgart.

Agenda 21 (1992) UN Department of Economic and Social Affairs, Division for Sustainable Development. www.un.org/esa/sustdev/agenda21.htm

Alexander E. B. (1988) Rates of soil formation: implications for soil loss tolerance. *Soil Sciences*, **145**, 37–45.

ALGLL (2004) Adjusting lime rates. Fact sheet 6. A & G Great Lakes Laboratories, Fort Wayne IN.

von Altrock C. (1995) Fuzzy logic and neurofuzzy applications explained. Prentice-Hall, New Jersey.

Aristotle 350 BC. On interpretation. Translated by EM Edghill.

Atterberg A. (1911) Über die physikalische Bodenuntersuchung und über die Plastizität der Tone (On physical soil investigation and on the plasticity of clays). Internationale Mitteilungen für Bodenkunde, **1**, 10–43.

Bagnold R. A. (1941/2005) The physics of blown sand and desert dunes. Dover Publications, 265p.

Barbault R. (2003) Écologie générale (General ecology). Dunod Paris, 326p.

Barrow C. J. (1994) Land degradation: Development and breakdown of terrestrial environments. Cambridge University Press, Cambridge, 313p.

Bartelme N. (2005) Geoinformatik (Geoinformatics). Springer, Berlin, Heidelberg, New York, 454p.

Barton N. (1973) Review of a new shear-strength criterion for rock joints. *Engineering Geology*, **7**, 287–332.

Barton N. (1986) Deformation phenomena in jointed rock. *Geotechnique*, **36**(2), 147–167.

Baumgardner M. F. (2000) Soil Databases. In M. E. Sumner (ed) *Handbook of soil sciences*. Taylor & Francis, Boca Raton: H1–18

BBR (2005) Spatial statistics of the Federal Office of Building and Regional Planning BBR, Bonn. www.bbr.bund.de

Bell F. G. (1993) Engineering geology. Blackwell, Oxford, 359p.

Bell F. G. (1993b) Engineering treatment of soils. E & FN Spon London, 302p.

Bell F. G., Genske D. D. and Stacey T. R. (2000) Mining subsidence and its effect on the environment: some differing examples. *Environmental Geology*, **40**, 135–152.

Bennett H. H. (1938) Soil conservation. McGraw-Hill, New York.

Berief K.-H. and Krupop F. (1997) Fundamente: dicke Brocken für die Folgenutzung (Foundations: stumbling stones for redevelopment). *BrachFlächenRecycling*, **4**(97), 14–18.

273

Bertemes F. and Schlosser W. (2004) Der Kreisgraben von Goseck und seine astronomischen Bezüge (The Goseck-circles and their relation to astronomy). In Meller H. (ed) Der geschmiedete Himmel. Wissenschaftliche Buchgesellschaft: 48–51.

Bezdek J. (1994) Fuzziness versus probability: again? In: IEEE Transactions on Fuzzy Systems, 2(1), 1–3.

Bichteler J. (1986) Expert systems for geoscience information processing. In: Proceedings of 3rd International Conference on Geoscience and Information. Glenside, Australia. 1, 179–191.

Blei W. (1981) Erkenntnisse zur Erd- und Lebensgeschichte: ein Abriss (Findings on the history of earth and life). WTB 219 Akademieverlag Berlin, 433p.

Block M. P. (ed) (1928) Der Gigant an der Ruhr (The giant on the Ruhr). Berlin.

Blossey S., Busch J., Dahlmann I., Feldwisch N., Oeser G. H., Penndorf O. and Schürer S. (2005) Entsiegelung von Böden im Rahmen der naturschutzrechtlichen Eingriffsregelung (Unsealing soil and ecological intervention standards in Germany). *Bodenschutz*, 2, 36–39.

Bolliger C., Höhener P., Hunkeler D., Häberli K. and Zeyer J. (1999) Intrinsic bioremediation of a petroleum hydrocarbon-contaminated aquifer and assessment of mineralization based on stable carbon isotopes. *Biodegradation* 10, 201–217.

Bork H. R. (1983) Die holozäne Relief- und Bodenentwicklung in Lösslehmgebieten: Beispiele aus dem südöstlichen Niedersachsen (The development of relief and soil in Holocene loess: Examples from southeastern Lower Saxony, Germany). *Catena* Supplement, 3, 1–93.

Bork H. R. (1989) The history of soil erosion in southern Lower Saxony. *Landschaftsgenese und Landschaftsökologie*, 16, 135–163.

Bork H. R., Bork H., Dalchow C., Faust B., Priorr H. P. and Schatz T. (1998) Landschaftsentwicklung in Mitteleuropa. Wirkungen des Menschen auf die Landschaften (Landscape development in Europe: The human impact on landscapes). Klett Gotha, 328p.

Bot A. J., Nachtergaele F. O. and Young A. (2000) Land resource potential and constraints at regional and country levels. Food and Agricuulture Organisation FAO Rome, World Soil Resources Report 90, 114p.

Bradshaw A. D. and Chadwick M. J. (1980) The restoration of land. Blackwell, Oxford.

Bradwell G. E. (1970) Some statistical features of the relatonship between Rocky Mountain Arsenal waste disposal and frequency of earthquakes. *Geological Society of America*, Engineering Geology Case Histories, 8, 33–37.

Brooks R. R. (ed) (1998) Plants that hyperaccumulate heavy metals. CAB International, Oxon, 380p.

Bruines P. (2003) Laminar ground water flow through stochastic channel networks in rock. PhD-Thesis ETH Lausanne, 129p.

Bruines P. and Genske D. D. (2001) Migration of contaminants in fractured media. Proceedings of the International Workshop on Migration of Contaminants in Fractured Media, 15th September 2000, Lausanne, Switzerland. 102p.

BFS (2002) Environment Switzerland 2002: Statistics and Analyses. BFS, Neuchâtel, 322p.

Buckley S. M. (2005) Mapping subsidence with satellite radar interferometry. Geological Society of America South-Central Section, 39th Annual Meeting, San Antonio, Texas.

Burt R. (ed) (2004) Soil survey laboratory methods manual. Soil Survey Investigation Report 42 (Version 4.0), Natural Resources Conservation Service, Lincoln, USA.

Carson R. (1962) Silent spring. Houghton Mifflin (reprinted 2000 in Penguin Classics, 323p).

Casagrande A. (1932) Research on the Atterberg Limits of soil. *Public Roads*, 13, 121–136.

Casagrande A. (1934) Die Aräometermethode zur Bestimmung der Kornverteilung (A method to characterise grain size distributions). Springer, Berlin.

Chapman H. D. (1965) Cation-exchange capacity. In: C. A. Black (ed) Methods of soil analysis: chemical and microbiological properties. *Agronomy*, **9**, 891–901.

Clark I. (1979) Practical geostatistics. Elsevier Applied Sciences, London, 129p.

Cowling E. B. (1982) An historical perspective on acid rain precipitation. In R. E. Johnson (ed) Proceedings of International Symposium on Acid Rain and Fishery Impacts on Northeastern North America. American Fisheries Society Special Publication, Bethesda MD, 15–31.

Cunningham S. D., Anderson T. A., Schwab A. P. and Hsu F. C. (1996) Phytoremediation of soils contaminated with organic pollutants. *Advances in Agronomy*, **56**, 55–95.

Darcy H. (1856) Les fountains publiques de la ville de Dijon (The public fountains of Dijon). Victor Dalmont, Paris.

Davis J. C. (2002) Statistics and data analysis in geology. John Wiley & Sons, New York, 638p.

Deere D. U. (1964) Technical description of rock cores for engineering purposes. *Rock Mechanics and Engineering Geology*, **1**(1), 17–22.

Diamond J. (2006) Collapse: how societies choose to fail or suceed. Penguin, 576p.

Dodt J., Genske D. D., Kappernagel T. and Noll P. (1993) Digital evaluation of contaminated sites. In: Engineering Foundation (ed) Digital image processing: techniques and application in civil engineering, Kona, Hawaii, 165–172.

Dosch F. (2003) Zahlen und Fakten zum Flächenverbrauch (Facts and figures of land consumption). VEGAS Symposium Rohstoff Fläche, Stuttgart.

Dotterweich M., Schmitt A., Schmidtchen G. and Bork H. R. (2003) Quantifying historical gully erosion in northern Bavaria. *Catena*, **50**, 135–150.

Drecker P., Genske D. D., Heinrich K. and Noll P. (1995) Subsidence and wetland development in the German Ruhr District. In: F. B. J. Bahrends, F. J. J. Brouwer and F. H. Schröder (eds) Land subsidence. IAHS Press, Wallingford, UK, 413–421.

Dupuit J. (1863) Etudes théoretiques et pratiques sur le mouvement des eaux (Theoretical and practical studies on the movement of water). Dunod, Paris.

Eastcott L., Shiu W. Y. and Mackay D. (1988) Environmentally relevant physical properties of hydrocarbon: A review of data and development of simple correlations. *Oil and Chemical Pollution*, **4**, 191–216.

Edwards A. J. (2004) ISO 14001 Environmental certification step by step. Butterworth-Heinemann, 272p.

EEA (1999) CORINE land cover: a key database for European integrated environmental assessment. European Environmental Agency Copenhagen.

EEA (2000) Down to earth: soil degradation and sustainable development in Europe—a challenge for the 21st century. European Environmental Agency Copenhagen.

EEA (2003) Europe's environment: the third assessment. Environmental assessment report 11, European Environmental Agency Copenhagen.

EEA (2006) www.eea.eu.int. European Environmental Agency Copenhagen (see information on 'Soil').

Ellerbrock R. (2000) Bestimmung der Kationenaustauschkapazität von Böden (Determining the CAC of soils). In H. Barsch, K. Billwitz and H. R. Bork (2000) Arbeitsmethoden der Physiogeographie und Geoökologie. Klett Gotha, Stuttgart, 334–337.

Ellison W. D. (1944) Studies of raindrop erosion. *Journal of Agricultural Engineering Research*, **25**, 131–181.

Elwell H. A. (1978) Modelling soil losses in Southern Africa. *Journal of Agricultural Engineering Research*, **23**, 111–127.

EMC (1998) Mont-Cenis. Report of the Entwicklungsgesellschaft Mont-Cenis, Herne D, 44p.

Emmett W. W. (1965) The virgil network: methods of measurements and a sample of data collected. *International Association of Scientific Hydrology Publication*, **66**, 89–106.

Engelen van V. W. P. (2000) SOTER: The world soils and terrain database. In M. E. Sumner (ed) *Handbook of soil sciences*. Taylor & Francis, Boca Raton: H19–27

Fairlie I. and Summer D. (2006) The other report on Chernobyl TORCH. The Greens EFA in the European Parliament. Brussels, 91p.

FAO (2000) Manual on integrated soil management and conservation practices: FAO land and water bulletin. Food and Agriculture Organization of the United Nations, FAO Rome.

FAO-AGL (2005) ProSoil: problem soils database. Food and Agriculture Organization of the United Nations, FAO Rome, www.fao.org

FAO-AGL (2000) Extent and cause of human induced soil degradation. Food and Agriculture Organization of the United Nations, FAO Rome, www.fao.org

FAO-Aquastat (2005) Information System on Water and Agriculture. Food and Agriculture Organization of the United Nations, FAO Rome, www.fao.org

Fielding E. J., Blom R. G. and Goldstein R. M. (1998) Rapid subsidence over oil fields measured by SAR interferometry. *Geophysical Research Letters*, **25**(17), 3215–3218.

Forter M. (2000) Bonfol oder Eine Lehrstück über Unrat (History of industrial waste handling in the City of Basel, Switzerland). *Die Weltwoche*, 20/2000

Fox F. A., Flanagan D. C., Wagner L. E. and Deer-Ascough U. (2001) WEPS and WEPP science commonality project. In J. C. Ascough and D. C. Flanagan (eds) Proceedings of the ASAE symposium on soil erosion research for the 21st century, 376–379.

Freundlich H. (1907) Über die Adsorption in Lösungen (On adsorption in solutions). *Zeitschrift für Physikalische Chemie*, **57**, 385–470.

FRTR (2006) Screening matrix and reference guide. Federal remediation technologies roundtable, www.frtr.gov/site

Fründ H. C. and Meuser H. (2000) Bodenfunktionen (Soil functions). WEKA Praxislösungen 2(5.3) Fachverlag für technische Führungskräfte, Kissingen, 1–82.

Fryrear D. W., Saleh A., Bilbro J. D., Schomberg H. M., Stout J. E. and Zobeck T. M. (1998) Revised Wind Erosion Equation RWEQ. Wind Erosion and Water Consevation Research Unit USDA-ARS, Southern Plains Area Cropping Systems Research Laboratory, Technical Bulletin 1.

Füchsel G. C. (1773) Entwurf zu der ältesten Erd- und Menschengeschichte, nebst einem Versuch, den Ursprung der Sprache zu finden (Outline of the oldest history of earth and man, including the attempt to find the source of language). Frankfurt/Leipzig.

Gantzer C. J., Anderson S. H., Thompson A. L. and Brown J. R. (1990) Estimating soil erosion after 100 years of cropping on Sanborn field. *Journal of Soil and Water Conservation*, **45**, 641–644.

Genske D. D. (2003) Urban Land: Degradation, Investigation, Remediation (with a foreword by Klaus Töpfer UNEP). Springer, Berlin, Heidelberg, New York, 333p.

Genske D. D. (2005) Ingenieurgeologie: Grundlagen und Anwendung (Engineering Geology: Principles and Application). Springer, Berlin, Heidelberg, New York, 588p.

Genske D. D., Gillich W., Kories H. and Olk C. (1992) Contaminated sites: data processing, visualization and interpretation International Conference on Geotechnics and Computers, Paris, Presses de l'École Nationale de Ponts et Chaussées Paris, 847–854.

Genske D. D. and Hess-Lüttich E. W. B. (1998) Zeit-Zeichen in der Geologie (Time-signs in geology). In E. W. B. Hess-Lüttich and B. Schlieben-Lange (eds) Signs and time/Zeit und Zeichen. Gunter Narr Verlag, Tübingen, 133–151

Genske D. D., Noll H. P. and Risse U. (1999) European strategies of rehabilitation of degraded land. In: I. Inyang and V. O. Ogunro (eds) Enviromental geotechnology and global sustainable development. CEEST, Uni Massachussetts at Lowell, 39–46.

Genske D. D., Heinrich K. and Hueb J. (eds) (2000) Strategy on sanitation for high risk communities: Identification and planning. Proceedings of an International Workshop, Lausanne. WHO, Geneva, Leylakitap-Verlag, Bern, Switzerland, 114p.

Genske D. D., Ruff A. (2006) Expanding cities, shrinking cities, sustainable cities: challenges, chances and examples. 10. IAEG-Congress Engineering Geology for Tomorrows' Cities, Nottingham.

Gerlach T. (1966) Wspólczesny rozwój stoków w dorzeczu górnego Grajcarka (Present development of slopes along the Grajcarek River). *Prace Geograf.* IG PAN 52.

Gillman G. P. (1979) A proposed method for the measurement of exchange properties of highly weathered soils. *Australian Journal of Soil Research*, **17**, 129–139.

Gillman G. P. and Sumpter E. A. (1986) Modifications to the compulsive exchange method for measuring exchange characteristics of soils. *Australian Journal of Soil Research*, **24**, 61–66.

Giono J. (1990) L'homme qui plantait des arbres (The man who planted trees). Gallimard Foliot Cadet Rouge.

Goudie A. (2000) The human impact on the natural environment. Blackwell, 528p.

Gray D. H. and Leiser A. T. (1982) Biotechnical slope protection and erosion control. Van Nostrand Reinhold, New York.

Grill B. (2004) Die tödliche ignoranz (The deadly ignorance). Die Zeit, Hamburg, **59**(30).

Gunn J. M. (ed) (1995) Restoration and recovery of an industrial region: the smelter-damaged landscape near Sudbury Canada. Springer, Berlin, Heidelberg, New York, 358p.

Hagen L. J. (1991) A wind erosion prediction system to meet user needs. *Journal of Soil and Water Conservation*, **46**, 106–111.

Hall P. (1999) Future urban lifestyles. In: BBR Bundesamt für Bauwesen und Raumordnung (ed.) Urban Future, Forschungen **92**, 31–40.

Hall P. and Pfeiffer U. (2000) Urban Future 21: A global agenda for twenty-first century cities. Proceedings of Urban 21 Conference, Berlin. Spon Press, Routledge, London.

Hannam I. and Boer B. (2004) Drafting legislation for sustainable soils: a guide. IUCN Environmental Policy and Law Paper 52, World Conservation Union.

Hanss M. (2005) Applied fuzzy arithmetic: an introduction with engineering applications. Springer, Berlin, Heidelberg, New York, 256p.

Hatheway A. W. (2002) Geoenvironmental protocol for site and waste characterization of former manufactured gas plants: worldwide remediation challenge in semi-volatile organic waste. *Engineering Geology* **64**, 317–338.

Hawking S. (2006) Life in the universe. Public Lecture. www.hawking.org.uk

Hazen A. (1892) Some physical properties of sands and gravels with special reference to their use in filtration. 24 Annual Report State Board of Health Mass Boston: 541–556.

Heim A. (1921) Geologie der Schweiz (Geology of Switzerland). Chr. Herm. Tauchnitz Leipzig, 1118p.

Heinrich K. (2000) Fuzzy assessment of contamination potentials. PhD Thesis TU Delft Netherlands, 208p.

Heitfeld K. H. (1979) Durchlässigkeitsuntersuchungen im Festgestein mittels WD-Testen (Permeability of rock mass measured with packer tests). *Mitteilung Ingenieur und Hydrogeoeologie Aachen* **9**, 175–218.

Herrick J. E. and Jones T. L. (2002) A dynamic cone penetrometer for measuring soil penetration resistance. *Soil Science Society of America Journal* **66**, 1320–1324.

Hoek E. and Bray J. W. (1981) Rock slope engineering. Institution of Mining and Metallurgy London, 358p.

Höhener P. (2001) Properties of LNAPLs and DNAPLs. Lecture notes. Swiss Federal Institute of Technology Lausanne EPFL, unpublished

Höhn C. (1996) Bevölkerungsvorausberechnung für die Welt, die EU-Mitgliedsländer und Deutschland. *Zeitschrift für Bevölkerungswissenschaft*, **21**, 171–218.

Hölder H. (1989) Kurze Geschichte der Geologie und der Paläontologie (Short history of geology and paleontology) Springer, Berlin, Heidelberg, New York, 244p.

Horn R. (2004a) Bearbeiten und Verdichten von Böden (Tillage and compaction of soils). In H. P. Blume (ed) Handbuch des Bodenschutzes. Ecomed-Verlag Landsberg am Lech, 188–219.

Horn R. (2004b) Sanierung verdichteter Böden (Remediation of compacted soils). In H. P. Blume (ed) Handbuch des Bodenschutzes. Ecomed-Verlag Landsberg am Lech, 801–805.

Horn R. and Fleige H. (2003) A method of assessing the impact of load on mechanical stability and on physical properties of soils. *Soil and Tillage Research Journal* **73**, 89–100.

Hudson N. W. (1964) Field measurements of accelerated soil erosion in localized areas. *Rhodesian Agricultural Journal* **31**, 46–48.

Hueb J. A. (2000) Preface in D. D. Genske, J. A. Hueb (eds.) High risk communities: identification and planning. International Workshop Lausanne, Swiss Federal Institute of Technology Lausanne EPFL and World Health Organisation WHO, Leylakitap Bern, 2–3.

Huth A. and Jürgens C. (1995) GIS- und Fernerkundungstechniken in der Bodenerosionsforschung (GIS- and remote sensing techniques in soil erosion research). *Geographische Rundschau* **47**, 450–456.

IPCC (2000) Land use, land use change and forestry. Intergovernmental Panel on Climate Change IPCC, Special Report (summary for policymakers www.ipcc.ch).

Isaaks E. H. and Srivastava R. M. (1989) Applied geostatistics. Oxford University Press, New York, 561p.

ISRIC (2005) GLASOD Global assessment of human-induced soil degradation. International soil reference and information centre ISRIC Wageningen, www.isric.org.

ISSS Working Group (1998) World reference base for soil resources: introduction and atlas. International Society of Soil Science (ISSS), International Soil Reference and Information Centre (ISRIC) and Food and Agriculture Organization of the United Nations (FAO). ACCO, Leuven, 165p, 79p.

IUGS (2005) Geoindicators: soil and sediment erosion. International Union of Geological Sciences www.iugs.org.

IUSS Working Group WRB (2006) World reference base for soil resources. World Soil Resources Reports 103, FAO Rome.

IWR (2005) RUSLE: The on-line soil erosion assessment tool. Michgan State University, Institute of Water Research. www.iwr.msu.edu.

Jacks G. V. and Whyte R. O. (1939) The rape of the earth: a world survey of soil erosion. Faber & Faber, London.

Jayawardane N. S., Blackwell J., Kirchhof G. and Muirhead W. A. (1995) Slotting – A deep tillage technique for ameliorating sodic, acid and other degraded subsoils and for land treatment of waste. In N. S. Jayawardane and B. A. Stewart (eds.) Subsoil Management Techniques. Advances in Soil Science, 109–146.

Jayawardane N. and Steward B. A. (1995) Subsoil management techniques. CRC, Boca Raton, 247p.

Jones R. J. A., Le Bissonnais Y., Dias J. S., Düwel O., Oygarden L., Bazzoffi P., Prasuhn V., Yordanov Y., Strauss P., Rydell B., Berenyi Uveges J., Loj G., Lane M. and

Vandekerckhove L. (2003) Nature and extent of soil erosion in Europe. Interim Report EU Working Group on Erosion WP2, Brussels, 27p.

Jürgens C. and Fander M. (1993) Soil erosion assessment by means of LANDSAT-TM and ancilliary digital data in relation to water quality. *Soil Technology* **6**, 215–223.

Kant I. (1784) Beantwortung der Frage: Was ist Aufklärung? (Answering the question: what is enlightenment?). Berlinische Monatsschrift **12**, 481–494.

Kant I. (1788, 1993) Critique of practical reason (Kritik der praktischen Vernunft). Prometheus, 193p.

Kayser A., Wenger K., Keller A., Attinger W., Felix H. R., Gupta S. K. and Schulin R. (2000) Enhancement of phytoextraction of Zn, Cd and Cu from calcareous soil: the use of NTA and sulphur amendments. *International Journal of Environmental Science and Technology*, **34**(9), 1778–1783.

Keller W. and Gunn J. M. (1995) Lake water quality, improvements and recovering aquatic communities. In J. M. Gunn (ed) Restoration and recovery of an industrial region. Springer, Berlin, Heidelberg, New York, 67–80.

Keren R. (2000) Salinity. In M. E. Sumner (ed) *Handbook of soil sciences*. Taylor & Francis, Boca Raton: G3–25

Kimble J. M., Lal R. and Grossman R. B. (1998) Alteration of soil properties caused by climate change. In: H. P. Blume, H. Enger, E. Fleischhauer, A. Hebel, C. Reij and K. G. Steiner (eds) Towards sustainable land use, Vol I, *Advances in Geoecology* **31**, Catena Reiskirchen, 175–184.

Klir G. J. and Folger T. A. (1988) Fuzzy sets, uncertainty and information. Prentice-Hall, London, 355p.

Kobori I. and Glantz M. H. (1998) Central Eurasian water crisis: Caspian, Aral and Dead Seas. United Nations University Press, Tokyo.

Krapp L. and Vorwerk S. (1997) Verfüllung teilweise verstürzter Mergelsteingrotten in Valkenburg/Niederlande (Injecting ancient limestone workings in Valkenburg, Netherlands), unpublished report.

Krey D. (1926) Erddruck, Erdwiderstand und Tragfähigkeit des Baugrunds (Earth pressure, earth resistance and bearing capacity). Ernst & Sohn, Berlin.

Krige D. G. (1951) A statistical approach to some basic mine valuations and allied problems on the Witwatersrand. *Journal of Chemical, Metallurgical and Mining Society of South Africa*, **52**(6), 119–139.

Kropla B. (2005) MapServer: Open source GIS development. Apress, 417p.

LaGrega M. D., Buckingham P. L. and Evens J. C. (2001) Hazardous waste management. McGraw-Hill, New York, 1184p.

Lal R. (1988) Soil erosion research methods. Soil and Water Conservation Society, Ankeny, IA, USA.

Lal R., ed (1994) Soil erosion and research methods. CRC Press, 352p.

Lal R. (2002) Soil conservation and restoration to sequester carbon and mitigate the greenhouse effect. In J. L. Rubio, R. P. C. Morgan, S. Asins and V. Andreu (eds) Man and soil at the third millennium. Geoforma Ediciones Logrono, 37–51.

Lane L. J. and Nearing M. A. eds. (1989) USDA water erosion prediction project: hillslope profile model documentation. NSERL 2 USDA-ARS National Soil Erosion Laboratory, West Lafayette

Lane J. R. and Perrott M. (1989) Eliza: remembering a Pittsburgh steel mill. Howell Press, Charlottesville, Virginia, 104p.

Lebert M., Brunotte J. and Sommer C. (2004) Ableitung von Kriterien zur Charakterisierung einer schädlichen Bodenveränderung, entstanden durch nutzungsbedingte Verdichtung von

Böden (Characterisation of soil damage due to soil compaction). Umweltbundesamt Texte 46–04 Dessau.

Lebert M. and Schäfer W. (2005) Verdichtungsgefährdung niedersächsischer Ackerböden (Agricultural soils susceptible to compaction.) *Bodenschutz* **2**, 42–46.

Lee L. S., Rao P. S. C. and Okuda I. (1992) Equilibrium partitioning of polycyclic aromatic-hydrocarbons from coal-tar into water. *Journal of Environmental Science and Technology*, **26**, 2110–2115.

Lessmann-Schoch U., Kahrer R. and Brümmer G. W. (1991) Pollenanalytische und C14-Untersuchungen zur Datierung der Kolluvienbildung in einer lössbedeckten Mittelgebirgslandschaft/Nördlicher Siebengebirgsrand (Dating colluvial deposits with pollen- and C14-analyses in the loess soils close to Bonn, Germany). *Eiszietalter und Gegenwart*, **41**, 16–25.

Lieberherr-Gardiol F. (1997) Waste, waste, nothing but waste. SKAT WasteNet Infopage, St Gallen, Switzerland.

Lillesand T. M. and Kiefer R. W. (2003) Remote sensing and image interpretation. John Wiley and Sons, 784p.

Long M. E. (2002) Half life: the lethal legacy of America's nuclear waste. National Geographic Magazine, July 2002.

Longley P. A., Goodchild M. F. and Maguire D. J. (2005) Geographic Information Systems and Science: John Wiley & Sons, 517p.

Losch S. and Cordsen E. (2004a) Bodenüberformung und Bodenversiegelung (Soil degradation and sealing). In H. P. Blume (ed) Handbuch des Bodenschutzes. Ecomed-Verlag Landsberg am Lech, 167–187.

Losch S. and Cordsen E. (2004b) Entsiegelungspotenziale (Potentials of unsealing). In H. P. Blume (ed) Handbuch des Bodenschutzes. Ecomed-Verlag Landsberg am Lech, 672–676.

Lugeon M. (1933) Barrages et Géologie (Dams and geology). Lausanne (Rouge) Paris (Dunod).

Lüttig G. (1960) Zur Gliederung des Aue-Lehms im Flussgebiet der Weser (Human-induced soils in the Weser Region, Germany). *Eiszeitalter und Gegenwart*, **11**, 39–50.

Lyell C. (1830-33) Principles of geology, being an attempt to explain the former changes of the earth's surface by reference to causes now in operation, **3**. John Murray, London.

McCutcheon S. C. and Schnoor J. L. (2003) Overview of phytotransformation and control of wastes. In S. C. McCutcheon and J. L. Schnoor (eds) Phytoremediation: transformation and control of contaminants. Wiley Intersciences.

Marsh G. P. (1864) Man and Nature. Scribner, New York.

Massonnet D., Holzer T. and Vadon H. (1997) Land subsidence caused by the East Mesa Geothermal Field, California, observing SAR interferometry. *Geophysical Research Letters*, **24**(8), 901–904.

Matheron G. (1962) Traité de Géostatistique Appliquée I (Applied geostatistics I). Technip Paris, 334p.

Matheron G. (1963) Traité de Géostatistique Appliquée II (Applied geostatistics II). Technip Paris, 172p.

Mathieu R, King C, Le Bissonnais Y (1997) Contribution of multi-temporal SPOT-data to the mapping of a soil erosion index: The case of the loamy plateaux of northern France. *Soil Technology* **10**: 99–110

Meadows D. H., Meadows D. L. and Randers J. (1972) The limits of growth. Universal Books, New York.

Meadows D. H., Meadows D. L. and Randers J. (1998) Beyond the limits. Earthscan, London, 300p.

Meadows D. H., Randers J. and Meadows D. L. (2004) The limits of growth: the 30-year update. Chelsea Green Publishing Company, 338p.

Meinel G. and Hernig A. (2005) Survey of soil sealing on the basis of the ATKIS basic DLM: feasibility and limits. In M. Schrenk (ed) 10th International Conference on Information and Communication Technologies (ICT) in Urban Planning and Spatial Development and Impacts of ICT on Physical Space. Proceedings CORP 2005 & Geomultimedia 05, Vienna, 359–363.

Meyer L. D. (1984) Evolution of the universal soil loss equation. *J Soil Water Conservation* **39**: 99–104

Meyer D. E. (1993) Flächen- und Stoffinanspruchnahme, Massenverlagerung (Occupation of land and matter). In H. Wiggering (Ed) Steinkohlenbergbau, Ernst & Son, Berlin, 116–121.

Meyer C. R., Wagner L. E., Yoder D. C. and Flanagan D. C. (2001) The Modular Soil Erosion System (MOSES). In J. C. Ascough, D. C. Flanagan (eds) Proceedings of the International Symposium of Soil Erosion Research for the 21st Century, Honolulu. American Association for Agricultural and Biological Engineers ASAE, St Joseph, MI, 358–361.

Montanarella L. (2003) Overview of soil resources and soil conservation in Central Eastern Europe. *Local Land and Soil News* **6** II: 5–6

Montgomery J. H. (1996) Groundwater Chemicals: Desk Reference. CRC Press Boca Raton Florida

Morgan R. P. C. (2001) A simple approach to soil loss prediction: a revised Morgan-Morgan-Finney model. *Catena* **44**: 302–322

Morgan R. P. C. (2005) Soil erosion and conservation. Blackwell, 304p

Morgan R. P. C., Morgan DDV, Finney HJ (1984) A predictive model for the assessment of soil erosion risk. *Journal of Agricultural Engineering Research* 30: 245-253

Morgan R. P. C., Quinton J. N., Smith R.E., Govers G., Poesen J. W. A., Auerswald K., Chischi G., Torri D., Styczen M. E. (1998) The European Soil Erosion Model EUROSEM: a dynamic approach for predicting sediment transport from fields and small catchments. *Earth Surface Processes and Landforms* **23**: 527–544

Mukaidono M. (2001) Fuzzy logic for beginners. World Scientific Publishing Company, 105p.

Mücher H. J. (1986) Aspects of loess and loess-derived slope deposits: an experimental and micromorphological approach. Amsterdam, 270p.

Myers N. (1988) Natural resource systems and human exploitation systems: physiobiotic and ecological linkages. World Bank policy planning and research staff, environmental department working paper, 12

Neteler M., Mitasova H. (2004) Open source GIS: a GRASS GIS approach. Kluwer Academic Publishers, 419p.

Nguyen T. Q., Helm D. C. (1995) Land subsidence due to groundwater withdrawal in Hanoi Vietnam. In: Bahrends F. B. J., Brouwer F. J. J., Schröder F. H. (eds) Land subsidence by fluid withdrawal and solid extraction. Balkema Rotterdam: 35–60

Oden S. (1970) Interaction between ocean and terrestrial ecosystems. In Singer D. S. (ed) Global effects of environmental pollution. Reidel Dordrecht.

Oldeman L. R., Hakkeling R. T. A., Sombroek W. G. (1991) World map on the status of human-induced soil degradation: an explanatory note. ISRIC Wageningen, 41p.

Or D. and Wraith J. M. (2000) Soil water content and water potential relationship. In M. E. Sumner (ed) *Handbook of soil sciences*. A53–85. Taylor & Francis, Boca Raton.

Ordonnance sur les sites contaminées (1998) Swiss Standard on Contaminated Sites, Switzerland.

Oswalt P. (2005) Schrumpfende Städte Band 2: Handlungskonzepte. Ostfildern, Hatje Cantz Verlag.

Oswalt, P. (2006) Shrinking cities. In D. Genske, M. Huch and B. Müller (eds.) Zukunft – Fläche – Raum. Schriftenreihe der Deutschen Geologischen Gesellschaft Hannover **37**: 203–207.

Oswalt P. and Rieniets T. (2006) Atlas der schrumpfenden Städte (Atlas of Shrinking Cities). Ostfildern, Hatje Cantz Verlag.

PADEP (2005) Pennsylvania stormwater best managment practices manual: section 6 (draft). Pennsylvania Department of Environmental Protection, 82p.

Pimentel D., Berger B. and Filiberto D. (2004) Water resources: agricultural and environmental issues. *Biosciences*, **54**(10), 909–918.

Press F., Siever R. and Grotzinger J. (2003) Understanding Earth. Freeman, 568 p.

Prinz H. (1997) Abriss der Ingenieurgeologie (Engineering Geology). Enke Stuttgart, 546p.

Proctor R. R. (1933) Four articles on the design and construction of rolled earth dams. Engineering News-Record, **111**, 245–248, 286–289, 348–351, 372–376.

Qureshi M. H. (2004) Ecological saints of deserts: their environmental and socio-economic profile. Rajat Publications. New Delhi, 172p.

Ramade F. (2005) Éléments d'écologie: Écologie appliquée (Elements of ecology: Applied Ecology). Dunod Paris, 864p.

Raven P. H., Berg L. R. and Johnson G. B. (1998) Environment. Saunders College Publishing, Fort Worth, 579p.

van Reeuwijk L. P. (2006) Procedures for soil analysis. Technical Report 9, ISRIC, Wageningen, Netherlands.

Renard K. G., Foster G. R., Weesies G. A. and Porter J. P. (1991) RUSLE: Revised universal soil loss equation. *Journal of Soil and Water Conservation*, **46**(1), 30–33.

Renard K. G., Foster G. R., Yoder D. C. and McCool D. K. (1994) RUSLE revisited: Status, questions, answers, and the future. *Journal of Soil and Water Conservation*, **49**(3), 213–220.

Renard K. G., Foster G. R., Weesies G. A., McCool D. K. and Yoder D. C. (1997) Predicting Soil Erosion by Water: A Guide to Conservation Planning with the Revised Universal Soil Loss Equation. US Department of Agriculture, Agriculture Handbook 703, 384p.

Reeves R. D. and Baker A. J. M. (2000) Metal-accumulating plants. In: I. Raskin and B. D. Ensley (eds) Phytoremediation of toxic metals using plants to clean up the environment. Wiley, New York, 193–230.

Rey F. and Nicot F. X. (2004) Ecological rehabilitation for erosion control in the Grands Goulet natural site, Vercors, France. *Local Land and Soil News*, **10/11**, 24.

Richter G, (1976) Bodenerosion in Mitteleuropa (Soil erosion in Central Europe). Wege der Forschung, 430 Darmstadt.

Rippen G. (2002) Handbuch Umweltchemikalien (Handbook Environmental Chemicals). Hüthig Jehle Rehm (on CD).

Ripper P. (1992) Sanierung des Altölraffinerie-Standortes Pintsch-Öl Hanau (Clean-up of hydrocarbon contamination at the Pitsch site in Hanau, Germany). In V. Gossow (ed) Altlastensanierung. Bauverlag Wiesbaden Berlin, 245–259.

Rokos D. and Kolokoussis P. (2004) Natural regeneration potential and soil erosion risk after forest fires in typical Mediterranean areas. *Local Land and Soil News*, **10/11**, II/III, 41–43.

Rose C. W., Williams J. R., Sander G. C. and Barry D. A. (1983) A mathematical model of soil erosion and deposition process. *Soil Science Society of America Journal*, **47**, 991–995.

Ross T. (2004) Fuzzy logic with engineering applications. John Wiley & Sons, 650p.

Ruff A., Stuth N. and Bierig R. (2006) Ein Bewertungsmodell für Flächenressourcen (An evaluation tool for land resources). In D. D. Genske, M. Huch, B. Müller (eds) Zukunft-

Fläche-Raum. Schriftenreihe der Deutschen Geologischen Gesellschaft DGG Hannover, **37**: 91–100.

Sander B. (1930) Gefügekunde der Gesteine (Joint survey and rock). Springer, Wien, 352p.

Sander B. (1948/1950) Einführung in die Gefügekunde der geologischen Körper (Introduction to structural geology). Springer, Wien and Innsbruck.

Schad H. and Teutsch G. (1999) Reaktive Wände: Aktueller Stand der Praxisanwendung (Reactive walls: current trends and examples). Flächenrecycling Brownfields Redevelopment **2** (99): 24-31

Scheibe R., Seidel K. and Zenk J. (1997) Geophysikalische Vorerkundung von unterirdischen Hindernissen bei Abbruch- und Erdarbeiten (Geophysical exploration of underground obstacles for clearance and remediation works). *Flächenrecycling Brownfields Redevelopment* **2**, 47–52.

Schiechtl H. M. (1991) Böschungssicherung mit ingenieurbiologischen Bauweisen (Slope stabilisation with bioengineering). In U. Smoltczyk (ed) Grundbautaschenbuch II , Ernst & Sohn, Berlin.

Schmidt J. (1991) A mathematical model to simulate rainfall erosion. In H. R. Bork, J. De Ploey and A. P. Schick (eds) Erosion, transport and deposition processes: theory and models. *Catena* Supplement **19**, 101–109.

Schmidt J. (1998) Modellbildung und Prognose zur Wassererosion (Models and prognosis for water erosion). In G. Richter (ed) Bodenerosion: Analyses und Bilanz eines Umweltproblems. Wissenschaftliche Buchgesellschaft Darmstadt, 137–151.

Schmidt R. G. (1983) Technische und methodische Probleme von Feldmethoden der Bodenerosionsforschung (Technical and methodological problems with field methods in erosion research). *Geomethodica* Basel **8**, 51–85.

Schmidt W. (1932) Tektonik und Verformungslehre (Tectonics and deformation). Bornträger, Berlin.

Schokking F. (1995) Prediction of long-term subsidence in drained peat areas in the Province of Friesland. In F. B. J. Barends, F. J. J. Brouwer and F. H. Schröder (ed.) Land subsidence: natural causes, measuring techniques, the Groningen Gasfields. Balkema, Rotterdam, 375–387.

Schubert H. (1999) Impact of Changes in Age and Household Structures on the Cities of the World. In: BBR Bundesamt für Bauwesen und Raumordnung (ed.) Urban Future, Forschungen **92**, 1–18.

Schumacher B. A. (2002) Methods for the determination of total organic carbon (TOC) in soils and sediments. US EPA Ecological Risk Support Center NCEA-C-1282, 25p.

Schwarzenbach R. P., Gschwend P. M. and Imboden D. M. (2002) Environmental organic chemistry. Wilcy, New York, 1314p.

Seiler W. (1980) Meßeinrichtung zur quantitativen Bestimmung des Geoökofaktors Bodenerosion in der topologischen Dimension auf Ackerflächen im Schweizer Jura (Device to measure soil erosion). *Catena*, **7** 2/3, 233–250.

Senatsverwaltung für Stadtentwicklung Berlin (2006) Digitaler Umweltatlas (Digital enivonmental map). www.stadtentwicklung.berlin.de.

Shao Y. (2000) Physics and modelling of wind erosion. Kluwer Academic Publishers, 389p.

Shiu W. Y., Bobra M., Bobra A. M., Maijanen A., Sunito L. R. and Mackay D. (1990) The water solubility of crude oils and petroleum products. *Oil and Chemical Pollution*, **7**, 57–84.

Sichardt W. (1928) Das Fassungsvermögen von Rohrbrunnen und seine Bedeutung für die Grundwasserabsenkung, insbesondere für größere Absenkungstiefe (Well capacities and their meaning for the lowering of the groundwater table). Springer, Berlin, Heidelberg, New York, 86p.

Skidmore E. L. and Williams J. P. (1991) Modified EPIC wind erosion model. Modelling plant and soil systems ASA-CSSA-SSSA Agronomy Monograph Madison WI **31**, 457–469.

Spencer C. H. (1997) Tutorial: the electromagnetic spectrum and remote sensing. *Earth Observation Magazine*, November 1997, 18–20.

Spies C. D. and Harms C. L. (2005) Soil acidity and liming of Indiana soils. Purdue University Department of Agronomy, AY-267-W, www.ces.purdue.edu.

Starr R. C. and Cherry J. A. (1994) In-situ remediation of contaminated ground water: the funnel-and-gate system. *Ground Water*, **32**(30), 465–476.

Stellman J. M., Stellman S. D., Christian R., Weber T. and Tomasallo C. (2003) The extent and patterns of usage of Agent Orange and other herbicides in Viet Nam. *Nature*, **422**, 681–687.

Stini J. (1922) Technische Geologie (Technical geology). Enke Stuttgart 789p.

Stockle C. O. (2001) Environmental impact of irrigation: a review. Washington Water Research Centre, Washington State University, 15p.

Stojek B. (2004) Plough sole as a result of the agricultural land use. *Miscellanea Geographica Warszawa*, **11**, 63–69.

Stolbovoi V. and Fischer G. (1997) A new digital georeferenced database of soil degradation in Russia. International Institute for Applied Systems Analysis IIASA Laenburg Interim Report IR-97-084, 16p.

Sunday Times (2000) A new danger hits 500 000 house prices. 17 September 2000.

Szelag S. and Weber U. (1993) Bergsenkungen (Subsidence). In: H. Wiggerung (ed) Steinkohlebergbau. Ernst & Son, Berlin, 121–135.

Tait M. and Hunter A. (2004) Modeling of anticipated subsidence due to gas extraction using kriging on sparse data sets. INGEO 2004 and FIG Regional Central and Eastern Conference on Engineering Surveying, Bratislava, Slovakia, 13p.

Terzaghi K. (1925) Erdbaumechanik auf bodenphysikalischer Grundlage (Foundation engineering based on soil physics). Deuticke, Leipzig/Wien.

Terzaghi K. and Peck R. B. (1967) Soil mechanics in engineering practice. 729 S. Wiley, New York.

Thiem G. (1906) Hydrogeologische Methoden (Hydrogeological methods). Gebhardt's Verlag Leipzig, 56p.

Toy T. J. and Foster G. R. (eds) (1998) Guidelines for the use of the Revised Universal Soil Loss Equation (RUSLE) on mined lands, construction sites, and reclaimed lands. USDI-Office of Surface Mining. Denver.

Trautner A., Fleige H., Van den Akker J. H., Arvidsson J., Horn R. and Pedersen K. (2000) Structure, development and use of a database about soil physical and mechanical properties and crop response as related to subsoil compaction. *Advances in GeoEcology*, **32**, 169–175.

Trümpy R. (1991) The Glarus Nappes: A Controversy of a Century Ago. In D. W. Mueller, J. A. McKenzie, H. Weissert (eds) Controversies in Modern Geology, Academic Press London, 385–404.

Tsimba R., Hussein J. and Ndlovu L. R. (1999) Relationship between depth of tillage and soil physical characteristics of sites farmed by smallholders in Mutoko and Chinyika in Zimbabwe. In P. G. Kaumbutho and T. E. Simalenga (eds) Conservation tillage with animal traction. Animal Traction Network for Eastern and Southern Africa ATNESA, Harare, Zimbabwe, 84–88.

UN-Earthwatch (2005) United Nations system-wide earthwatch. United Nations, earthwatch.unep.ch.

UNEP (2002) Change and challange: A state of the environment biefing for the global environmental facility. United Nations Environmental Programme Nairobi, 26p.

UNEP (2003) Global environmental outlook. 3: fact sheet. United Nations Environment Programme Nairobi, 4p.

UNEP (2004) North East Asian dust and sand storms growing in scale and intensity. Press release of the United Nations Environment Programme (8th Special Session of the Governing Council/Global Ministerial Environment Forum), Nairobi.

UNEP (2005) Geo year book: An overview of our changing environment. United Nations Environmental Programme Nairopi, 104p.

UNEP-GRID (2004) The disappearing Aral Sea. The Environment Times, UNEP-GRID Arendal, www.environmenttimes.net.

UNEP/UNCHS (1999) The Kosovo conflict: consequences for the environment and human settlings. United Nations Environmental Programme UNEP and the United Nations Centre for Human Settlement Habitat, UNEP Nairobi, 108p.

Unger P. W. and Kasper T. C. (1994) Soil compaction and root growth: A review. *Journal of Agronomy and Crop Science*, **86**, 759–766.

UN-Habitat (2003) The challenge of slums. Earthscan.

USGS (2000) Delta subsidence in California: The shrinking heart of the state. USGS Fact Sheet 005-00, United States Geological Service.

Vidal M. (1966) La terre armée (Reinforced soil). Annales de l'Institute Technique du Batiment et de Travaut Publiques, France, 888–938.

Vogelsang D. (1997) Environmental geophysics: a practical guide. Springer, Berlin, Heidelberg, New York, 173p.

Wackernagel M. and Rees W. (1996) Our Ecological Footprint: Reducing Human Impact on the Earth. New Society Publishers, Gabriola Island BC.

von Weizsäcker E. U. (2006) Antwort zu Stephen Hawkings Frage "Wie kann die Menschheit die nächsten hundert Jahre überleben?" (Answer to Stephen Hawking's question "How can the human race survive the next hundred years?") *Die Zeit*, **34**, 49

von Weizsäcker E. U., Lovins A. B. and Hunter L. (1997) Factor four: doubling wealth halving resource use. Earthscan, London, 322p.

Whitelaw K. (2004) ISO 14001 environmental systems handbook. Butterworth-Heinemann, 272p.

Wildhagen H. and Meyer B. (1972) Holozäne Bodenentwicklung, Sedimentbildung und Geomorphogenese im Flussauenbereich des Göttinger Leintal-Grabens (Human-induced Holocene soils in the Göttingen-Leintal Region, Germany). Göttinger Bodenkundliche Berichte **21**, 1–158.

Williams R. H. (1995) Contaminated land: a problem for Europe? *Flächenrecycling Brownfields Redevelopment*, **1**(95), 19–24.

Wilson S. and Haines S. (2005) Site investigation and monitoring for ground gas assessemnt: back to basics. *Land Contamination and Reclamation*, **13**(3), 211–222.

Winterhalder K. (1995) Early history of human activities in the Sudbury area and ecological damage to the landscape. In J. M. Gunn (ed) Restoration and recovery of an industrial region. Springer, Berlin, Heidelberg, New York, 17–31.

Wischmeier W. H. and Smith D. D. (1965) Predicting rainfall-erosion losses from cropland east of the Rocky Mountains. Agriculture Handbook 282, US Dept of Agriculture, Washington DC.

Wischmeier W. H. and Smith D. D. (1978) Predicting rainfall erosion losses: A guide to conservation planning. Agriculture Handbook 537, US Dept of Agriculture, Washington DC.

Woodruff N. P. and Siddoway F. H. (1965) A wind erosion equation. *Proceedings of Soil Science Society of America*, **29**, 602–608.

Wolman M. G. and Schick A. P. (1967) Effects of construction on fluvial sediment: urban and suburban areas of Maryland. *Water Resources Research*, **3**, 451–464.

World Commission on Environment and Development (1987) Our common future (Brundtland Report). Oxford University Press, Oxford.

WWF (2004) Living planet report 2004. World Wide Fund for Nature WWF, Gland, Switzerland. 42p.

Young A. (1998) Land resources: now and for the future. Cambridge University Press, Cambridge.

Zadeh L. A. (1965) Fuzzy sets. *Information and Control*, **8**, 338–353.

Zech W. and Hintermaier-Erhard G. (2002) Böden der Welt (Soils of the world). Spektrum Akademischer Verlag Heidelberg Berlin, 120p.

Index